Principles
Spray and Compost Preparations

Manfred Klett

Principles of Biodynamic Spray and Compost Preparations

Floris Books

First published by the International Biodynamic Initiatives Group
as *The Biodynamic Spray Preparations* in 1994
and as *The Biodynamic Compost Preparations* in 1996.
This revised edition published in 2006 by Floris Books

British Library CIP Data available

ISBN-10 0-86315-542-1
ISBN-13 978-086315-542-0

Produced in Poland by Polskabook

Contents

Foreword

The preparations are central to the biodynamic approach to farming and gardening and they have been used to good effect on farms and gardens across the world. Much has been written and spoken about them over the years and yet there is always more to learn. The biodynamic preparations bring questions with them that demand a lifetime and more of inner work and study to answer.

This may seem puzzling to the general reader and yet it is the struggle to understand how the preparations work that leads to a deepened understanding for life, nature and the human being, and which in turn supports the creative and intuitive capacities needed to farm the land. All too often we limit ourselves to the outer appearance of things and processes instead of entering into their deeper and wider context, and appreciating the subtle influences streaming in from the wide expanse of the heavens or acknowledging the spirituality that lives behind our physical existence. These are all key elements for understanding biodynamic agriculture.

The content of this book was originally given in the form of lectures during two biodynamic conferences held in Britain between 1992 and 1994. These International Biodynamic Initiative Group (IBIG) conferences were led for many years by Manfred Klett. At the time he was the Leader of the Agriculture Section of the School of Spiritual Science at the Goetheanum, Dornach, Switzerland. Prior to this he was a founding member of the Dottenfelderhof farm community near Frankfurt in Germany and farmed there for twenty-one years.

The first of the IBIG conferences took place in 1986. An important objective was to work with the fundamental ideas of biodynamic agriculture in an English speaking context. Manfred Klett was invited to do this as a keynote speaker. He did this for about ten years and brought with him his considerable experience and far reaching insights into biodynamic agriculture and spiritual science.

The lectures given in the early conferences addressed many different aspects of biodynamic farming and indeed agriculture as a whole. They contain gems of wisdom and are full of inspiration. They continue to be available as individual conference booklets (see Bibliography). The lectures on the preparations, reprinted here, were given over a three year period.

Beginning with a great sweep through the historical development of human consciousness, Manfred Klett describes the cultural context of agriculture before going on to explore the materials, substances and processes of the various preparations in more detail.

The subject matter is addressed in a very practical down to earth manner and with a deeply sensitive appreciation for life in agriculture. At the same time his thorough knowledge of anthroposophy radiates throughout. The book will provide lots of inspiration for anyone wishing to study the biodynamic preparations in more depth.

Bernard Jarman
November 2005

1. The Development of Consciousness and of Agriculture

In order to understand the spray preparations, one has to come to an understanding of the whole mystery of manuring. When he returned from Koberwitz (where he gave the series of lectures for farmers known as the Agriculture Course), Rudolf Steiner said in a lecture in Dornach that the mystery of manuring cannot be understood by natural science.[1] It can only be understood by those who are able to reach with their thoughts into the spiritual worlds. Only through spiritual research will we be able to understand the secrets of manuring. When manuring, we work with physical matter, but its inner nature is not revealed and therefore we work with a mystery. In order to form a basis for our theme I would like to look at the history of manuring, at the essence of its progression throughout its development in history.

Manuring is very closely connected to the development of human consciousness. The pre-Christian eastern tradition designated the past into 'ages.' There was the Golden Age, which was a very long-lasting age. During this time humankind was not yet fully incarnated. It was still embedded in the spiritual world. This age was followed by the Silver Age, and this by the Age of Ore. Their duration became ever shorter. The division of these ages can be a picture of how the human being has descended from the spiritual down to the physical world. The Age of Ore was followed by Kali Yuga or the Age of Darkness. Rudolf Steiner refers to this in the Agriculture Course. That is why I am introducing it here. He mentions the end of the Age of Darkness and gives the date as the year 1899. It lasted for five thousand years, and began in the year 3102 BC. We are now in the Age of

Light, although sometimes it seems even darker than the previous age! But where there is much light there is also much shadow.

Past ages

The development of these ages broadly corresponds to a cosmic rhythm. It is revealed when we consider the sunrise at the vernal equinox, on March 21. Behind the sun the zodiacal constellation of the Fishes, Pisces, appears rising. Going back in history, say to the Roman times, the sun did not rise in this constellation but in the one next to it, in Aries. The sun goes on a backward precession through the signs of the zodiac, and this journey takes about 25,920 years to complete: the Platonic Year. If we divide this figure of 25,920 years by twelve (representing the twelve signs of the zodiac) we get 2160 years. This is the average time it takes for the vernal point to pass through one sign of the zodiac. This rhythm of 25,920 is also found within the human being — the heartbeat and breathing relate to it.

The same rhythm is imprinted into the whole of humanity and is followed by the development of human consciousness. Each cultural epoch lasts approximately 2160 years. Anthroposophy defines a cultural epoch as the period in which a step in the development of consciousness is achieved, and this lasts about 2160 years. Rudolf Steiner indicated the year 1413 as the beginning of our present epoch, of the development of the consciousness soul. Dating back 2160 years we come to the year 747 BC which is normally cited as the founding of Rome, the beginning of the Greco-Roman epoch. If we again go back 2160 years we come to the year 2907 BC. This comes to the time of the Egyptian/Chaldean/Babylonian epoch. Another 2160 years back, we reach the year 5067 BC which can be determined as the beginning of the Ancient Persian culture. The preceding Ancient Indian epoch has its beginning in the eighth millennium BC. That is the time of the beginning of the Ancient Indian epoch. This date is the beginning of the post-

Atlantean development of humankind. Prior to this anthroposophy speaks of the Atlantean period, which was also divided into seven epochs during which humanity was endowed with the ego and out of its forces was developing the physical body.

The post-Atlantean period began with the Ancient Indian epoch, where the ether body developed in connection with the ego. During this Ancient Indian epoch the physical and etheric bodies of people were like an instrument upon which the spiritual hierarchies played. The last remnants of this harmony with the spiritual spheres appears in the Vedas.

Ancient Persia

The next step was taken during the Ancient Persian culture, when humankind really 'arrived' on earth, and opened the senses to perceive the outside world. This was the time when the astral body was formed out of the ego forces. The spiritual leader at this time, Zarathustra, taught his people to look at the physical world and to discover nature as the 'clothes' of the spiritual hierarchies. The earthly realm can be understood as a mantle of the spiritual beings which has to be penetrated in order to come to a new understanding of the creative beings. He also taught of the polarity of heaven and earth. In heaven, the realm of light, Ahura Mazdao rules, while the darkness is ruled by Ahriman. He gave the prophecy that one day the mighty Sun Being would be incarnated amongst humanity. Guided by the teachings of Zarathustra, people started to cultivate the earth, to rip it up with a plough. Mythologically the first plough was looked on as a golden one. This signifies that it was a sun impulse to cultivate the soil, and in so doing to fertilize the earth. This can be seen as the first step of manuring. The earth was ripped up in order that warmth, air, and moisture might penetrate it. Through the introduction of warmth and air, we say today, a decomposition of humus takes place. The natural fertility of the soil is broken down and re-established by soil tillage and plant cultivation. In that ancient time there was the awareness that a change in death and life processes is a preliminary to fruit formation. It was at this time and during the following pre-Christian ages that all our food plants were first bred. This outstanding cultural event was closely related to 'manuring' through soil cultivation.

Moving on to the next epoch we enter the beginnings of the Age of Darkness and the time when the development of the human soul began. This could only take place through the human being disconnecting more and more from the spiritual world. This is why it was called the Age of Darkness, because the light of the spiritual worlds that once shone upon humankind had diminished, and people were left to themselves. The first stage of individualizing the soul by the ego forces was achieved by the development of the sentient soul.

The Egyptian epoch is essentially characterized by this withdrawal of the spiritual world. People were deeply connected to death. Only in the centres of mystery wisdom were the priest-kings able to mediate between the spiritual world and the social body. Rudolf Steiner indicates that in the Egyptian culture, the world of the priest had such an immense moral power that the whole social body acted according to it. But this magic force eventually vanished. The major agricultural settlements at that time were in the river valleys of the Euphrates, Tigris and Nile. The cultivated areas were regularly flooded by the river and people were able to control this flooding, and irrigate the soil. In the ancient Persian culture agriculture had developed in a quite different area, most probably in the area of the Hindu Kush where many of the wild species of our food plants are still to be found, and in the adjacent plains between Oxus (Amu-Darya) and Jaxartis (Syr-Darya) where there is a very fertile soil. They only had to plough it up in order to release the fertility. In the Egyptian culture agriculture took place in the valleys, which were fertilized by natural flooding and man-made irrigation. They were fertilizing not so much the earth but the plants with water and its loamy deposits. This is a very important distinction.

The Greco-Roman Age

Moving on to the Ancient Greek culture, the development of the intellectual soul took place. Think, for instance, of how Greek philosophy developed over this time. The ego took hold of the ether body and transformed it into the intellectual soul. This is the time when humankind arrived fully on earth, and was completely disconnected from the spiritual world. For the first time people felt death as an inner experience. They said 'Rather be a beggar in the upper world than a king in the realm of the shadows.' They were deeply concerned with the mystery of death, and there was a tremendous will to enliven what was dead. They took quarried marble and chiselled it into wonderful sculptures, in order to enliven and shape it in such a way as to give it the shine of life. But then, in the abyss of the Age of Darkness, the Mystery of Golgotha took place. It was the event that was to overcome death in the material world. The mighty being, to whom Zarathustra had referred, incarnated into a human body, and spiritualized it to the extent that earthly matter could participate in resurrection, and once again find its connection to its spiritual origin.

During the Greco-Roman epoch, agricultural development reflected the activity of the intellectual soul in well-organized gardens, enclosed fruit trees, growing on terraces and well-structured landscapes. The result of human thought reflects itself onto the outer world. The methods of cultivation adopted at this time came from an intellectual approach, for example the pruning of fruit trees.

Europe

Continuing on through history, we enter the post-Christian times, and the wonderful transformation that took place since the eighth and ninth centuries after Christ, which was the birth hour of European agriculture.[2] What later became the organism of the farm had its beginnings with the church in the centre of the village, encircled by the farm houses, animal husbandry, farm gardens, orchards and meadows, and arable crops which were all integrated as one organism. Since that time, which was still the time of the development of the now Christianized intellectual soul, people started to manure the earth itself, to enliven the mineral realm. Before the time of Christ the highest cultural approach was to shape the mineral kingdom into sculptures. In the preceding Egyptian culture all fertilizing was based on water but now the crucial step towards enlivening the earth was taken. Since the eighth and ninth centuries it became the common custom to relate animal husbandry to the arable work. Thus, cultivated land benefited from the addition of cow manure. It is very strange that cow manure, with such an outstanding fertilizing force, was not used for manuring in pre-Christian times. Even today in Africa and India it is left to dry in the sun and burned for fuel.

Since these early medieval times cattle lived very close to the people, even under one roof, and the farm buildings were all around the church, and the farmyard manure became the 'gold' of the farmer, so the old saying goes. The development of agriculture at that time is characterized by the animal husbandry methods adopted, which form the core. This farm organism has been developing in middle and eastern Europe right up until the middle of this century. But in the early Middle Ages the consciousness soul had not yet incarnated, and people lived in a dreaming consciousness and were guided by their own folk spirits. So, since the beginning of the new age which began in 1413, each country developed particular characteristics. The English folk instinctively developed the consciousness soul, the French developed the intellectual soul instinctively, the Italians the sentient soul and so on. The people awoke more and more and the individual ego began to guide all the activities of daily life. The connection to the spiritual world was completely severed. Only for the most advanced of that time did the ego act as an individual guide. Let us think, for example, of Nicolas Copernicus (1473–1543) or Galileo Galilei (1564–1642) or Francis Bacon (1561–1626). It was Francis Bacon who closed the doors on the spiritual worlds completely. He said we only need to rely on what we can see in the physical world around us and draw the laws from it by intellectual reflection. Newton and others continued with this approach.

End of the Dark Ages

Then we come to the year 1899. This marks a very crucial threshold. The closer we come to it the more materialistic the world became. The consciousness soul had taken hold of the physical and was eagerly trying to reveal its forces. It was in the year 1879 that the guiding archangel, Michael, ascended to become a time spirit. This caused dramatic changes. On the one hand, materialism developed to its utmost conceptually but at the same time, as a result of an intense experience of loneliness, freedom worked its way to the surface for human beings. This striving reached its culmination in the *Philosophy of Spiritual Activity* written by Rudolf Steiner in 1894. This book provides proof that the human being is free in his deeds if he has knowledge of his motives. This work was written for all humankind just before the end of Kali Yuga, and is really a summary of the preceding five thousand years of the development of the soul. In the year 1899 it is as if there were a signpost on which two signs pointed in different directions.

From this time onwards, human freedom could be achieved. But the consequence of gaining freedom is to be entirely alone and disconnected from one's spiritual origin and fellow human beings. But if you are not conscious of this fact you may go the way of arbitrariness. The consciousness soul thus leads initially to egotism, and develops anti-social forces. The whole of modern civilization has taken this path. We are now suffering the effects of it and wherever we look we see signs of a breakdown, of decadence. Tracing this path we see modern agriculture which, since the nineteenth century, has developed a new technology in fertilizers by synthesizing an earthly substance, a salt, out of the air. But this nitrogen salt is not really an earthly substance because once you apply it to the earth, it dissolves in water and activates it. Therefore it displays watery, not earthly, qualities. When using nitrogen fertilizer we once again fertilize the plant like in pre-Christian times. We do not enliven the earth. In those former times fertilizing by water was guided by the mystery centres, while now it is guided merely by our egotism. We apply a salt which dissolves in water and nourishes the plant and all this can be done without the need for solid earth at all (for instance with hydroponics). So what happens is nothing but an untransformed repetition of the Egyptian culture. There is a principle law of evolution: whenever something continues on, untransformed, into a later time, it becomes an evil. What has been good in its proper time becomes evil if it remains untransformed.

That is one of the paths we look at nowadays which has led to our present environmental crisis. The other path originates in the *Philosophy of Spiritual Activity.* Out of his inner activity, Rudolf Steiner went on to develop research into the spiritual world in a similar way to and in full acknowledgement of research into the natural world. This spiritual research, called anthroposophy, connects the human being to his spiritual origin. One could say that anthroposophy, founded in 1902, takes up the thread of the Ancient Persian culture. Before Kali Yuga began Zarathustra taught his pupils to find a new relationship to the spiritual world by using their senses to seek it through the natural world. We have now passed Kali Yuga, and the awakened ego in the consciousness soul is now prepared to embark upon the path of knowledge of higher worlds.

Rudolf Steiner's Agriculture Course

Towards the end of his teachings in anthroposophy, in 1924, Rudolf Steiner gave the Agriculture Course. There he revealed the guiding idea for future agriculture, to conceive a farm in its true being as a kind of self-contained individuality. He characterizes this fundamental idea as the threefold human being upside down.

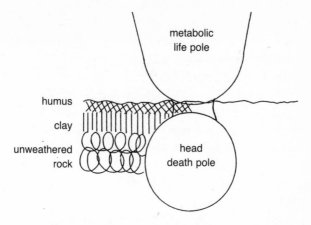

The head represents the mineral solid earth, and the metabolism is active in the realm above the earth, while the soil, related to the diaphragm, is the mediator between these two poles.[3]

The starting point of the Agriculture Course is opposite to Zarathustra's. He taught out of the consciousness of the macro-cosm, and directed his pupils towards a perception of the sentient world. But we have completely dislocated ourselves from this knowledge and taken hold of the physical world. Now we have to find, passing through the gateway of freedom out of our own awakened ego, a perception of the two poles beneath and above the earth. The Agriculture Course therefore starts with the microcosm, the human being. We first have to behold ourselves as human beings in order to understand the inner reality of the world. In the times of Zarathustra a knowledge of the outside world still pre-vailed and people were taught to perceive their own being out of this understanding. Not a repetition but a complete transformation, as a result of the Mystery of Golgotha, reappears in our time. The agricultural individuality is the central image we have to work with in biodynamic farming, to reach an understanding of the world by looking at ourselves. Anthroposophy provides the means to under-stand the human being, and our own spiritual origin. So we have to study anthroposophy as a whole in order to understand what we need to do in biodynamic farming. The Agriculture Course could not have been given by Rudolf Steiner in earlier years. The time was not yet ripe. It was only possible in 1924. This becomes obvious if we look again to the other path of development in our time. We see that modern science has almost achieved its aim to split the atom and thereby to release energy that could be capable of destroying the world. The Agriculture Course had to be given to provide a new view of things which balances out this destructive force. It contains indications of a technology in the living sphere: the use of the biodynamic preparations.

In modern conventional agriculture, there is no necessity for the farmers to create a personal relationship to their farms; it is a kind of press-button technology. I spoke to a conventional farmer not so long ago and he told me, 'In autumn, I calculate my harvest for then next year with my computer. And when I see what we actually harvest it is more or less accurate.' By this modern scientific means we are able to calculate the harvest even before the crops are sown, quite independently of all the natural conditions! So modern science transforms agriculture into a quantifiable, calculable business. Therefore there is no reason why the farmers should search for a new relationship between themselves as a microcosm and the inner nature of the macrocosm. But when one works with the indications given by Rudolf Steiner in the Agriculture Course, you will soon experience that successful farming requires the building up of a personal relationship. This is based on a sound scientific approach but leads further to practising an art.

Biodynamic farming in its essential being has to be discovered as a new art. This means we have to develop a personal relationship to bridge the abyss between our own spiritual being and the outside world. Whatever we practise in farming we should realize that it is at first hand a repetition of past achievements. If we build up a compost heap, for instance, we need to see that the decomposition and humus formation comes about in the right way. So in order to understand this process fully we should do it by hand now and then. It is not something that just happens out of itself, we extend our being into this process and thus build up a personal relationship. If we work with cow dung, this is even more the case. It is no good to just throw it somewhere in a heap, but we need to work with it consciously, throughout the year, to smell it, take it in our hands and observe the structure and colour. Of course we must appreciate the scientific analysis of how much nitrogen, potassium, etc, is in it but this must be complemented by feeling it and touching it and learning to perceive it as a substance that contains mighty forces.

This kind of relationship leads us to understand the biodynamic preparations on a new level. The preparations work into the living sphere of an organism. It is not a mechanism which you can control by pressing a button; an organism is a living being which is separated from its environment by having an inner life and sentient body. This inwardness underlies the laws of polarity. The most developed organism is that of the human being, and we can study these polarities by looking at ourselves. For example, we behold these polarities in sympathy and antipathy, in consciousness and unconsciousness, and in the nervous system and metabolism, with the rhythmical sphere in between. Our life is a tension between these polarities. Furthermore, the laws of transformation, development and life are followed in an organism. We must be clear that working with the biodynamic preparations means to work into the inner life of organisms to educate them, as it were. They work into the realm of intensification, between these polarities of transformation, and into the processes of dying and becoming. It is our task, now and in the future, to guide these processes, not arbitrarily, guided by our own egoism but freely, guided by anthroposophy.

The biodynamic preparations, especially the spray preparations, are the means by which we strengthen a living organism, to undergo these various processes. The farm as a whole is such an organism and in building up a new relationship by working with the preparations, we extend our being into the farm organism.

2. Substances of the Spray Preparations

If we try to discuss the question of how to make the spray preparations, and share our experiences, we will find that everybody has their own views! Doubtless this is an expression of our personal relationship to them. One person may say, 'Well, I think it is a very good thing to work with these spray preparations. They are an invention of anthroposophy. I've studied anthroposophy and can say I've found some logic in it, so there might be some logic in these spray preparations too, so I'll try them.' For this person it is a kind of belief, and I think everybody is at this stage where confidence in anthroposophy is the main part of understanding the preparations.

Another will come along and say, 'We want some scientific evidence! We believe in the preparations, but in order to understand them we must have some scientific proof.' But it is an illusion, to attempt to understand through scientific proof something that you believe in.

Others say, 'To spray the preparations is quite a challenge, and we have such difficulties doing the other work on our farm. Why don't we get a machine to do the stirring ... or we could use some Flowforms, which are so aesthetically pleasing when one watches the water moving in them.' So everybody starts to alter the process through their own opinion and judgment.

The personal relation to the preparations

There are two conditions we have to consider in order to build up a personal relationship. There is not one abstract understanding of the truth: 'Each has his own truth and in the end they all prove to be the same,' (Goethe). The first condition is that one needs to be very careful about one's own opinions, and to study exactly what recommendations are given. Actually, as biodynamic farmers, we ought to know the Agriculture Course by heart. Whatever work we do in our daily lives in the future should be permeated by this spiritual knowledge. With this you can judge your personal relationship, but no sooner. The other condition for building up a personal relationship is to actually do it! You have to make the preparations, stir them, apply them yourself. While working with them you must learn to converse within yourself with those words Rudolf Steiner put his spiritual findings into.

The two conditions to come to a deeper understanding lead us to follow a path of knowledge which is closely related to the path to 'the knowledge of the higher worlds' (reflecting the title of one of Rudolf Steiner's fundamental books). We have to take a step on this path, because the proof is not just an intellectual realization. We experience a kind of landscape, looking to the right and left, above and below. Walking through a physical landscape, the world speaks to us. So, too, we must open our hearts and minds while making the preparations, stirring them, spraying them, and beyond, and thus we will find the pictures that become imaginations that foster inner certainty in us. So I would like to take some steps along this path, and describe what can be observed in order to grasp the reality of the surrounding landscape.

Cow horns

At the outset, there are quite clear indications from Rudolf Steiner. You take cow horns and fill them in autumn with cow manure or with ground silica in the springtime. You bury them in the soil during winter, the other during summer, and take them out again in the spring or autumn, accordingly. You take a little of the contents and stir it rhythmically in water for one hour, and this highly diluted liquid is sprayed on the land. It is quite an easy procedure. Right from the beginning we have to try to meet these indications from

Shorthorn Cow

Rudolf Steiner with an open mind. Mere faith or antipathy prevent you from really experiencing them. Approaching them without prejudice we encounter a tremendous polarity, and thereby a discovery can be made. Polarities are always connected, and transformation and evolutionary development can take place. To give an example, think of the human being and how we are polarized between the head and the metabolism. These two, the thinking pole and the will pole contradict each other and yet are mutually necessary. It is from the point of view of polarities that I want to look at the spray preparations.

Let us first consider the horn/cow dung preparation. If we observe the horn and the cow dung we meet a polarity straight away! The horns are situated right on top of the skull of the cow, ideally pointing upwards. They form the highest point on the animal. The horns are a gift from the earth to the cow. She is not born with them, and it takes some time before they become visible. First, some time after the calf is born, it begins to lose a little hair to the right and left on the head. If you feel one of these patches with your thumb you can feel a loose, moving knobble of tissue

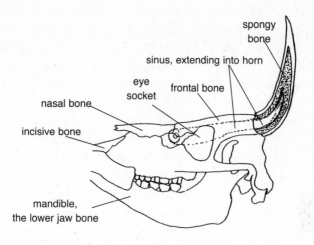

Longitudinal sketch of a cow's skull

underneath the skin. After some time this cartilaginous material becomes solid, while the skin grows over the top, which has become a round tip. If you look at a longitudinal sketch of the skull there is the nose bone, the frontal bone and the sinuses, which are caverns filled with air. There is the brain cavity. The skin has seven layers and the outermost layer, which is the hairs, start to form the horn. The horn is very densely formed hair over a bony substance which protrudes into the hollow of the horn. It develops after the calf is born, and is a gift of the earth.

It is interesting to see that the development of the horn is very closely related to the development of the whole digestive system, especially the rumen, which is also a gift from the earth to the calf. It only develops fully once the calf takes in earthly food. Another phenomenon is that there is a connection between the horn formation in the ruminants and the fact that they have no incisor teeth in the upper jaw. This was a discovery of Goethe, who made a special study of the intermaxillary bone, and its connection to teeth formation.[1] All the animals that develop incisors do not have the forces to develop horns. Once again, incisors relate to the whole metabolism. Another indication that the development of the horn is connected to the activity in the stomachs is that a ring can be seen on the cow's horn after the calf is born. One can reckon how many calves a cow has given birth to by counting these rings, which are formed by a slowing down in the growth of the horn at the onset of lactation. All these phenomena show the connection between the top of the skull and the metabolic system.

The circulatory system is also very much related to what takes place in the horn. The bone inside the horn is permeated with blood vessels (if the cow loses a horn then there is an immense loss of blood). There is a constant flow of blood into and out of the horn. The breathing system is also involved. When the cow inhales air, some goes up into the sinuses, which are very well developed in ruminants. The rhythms of respiration and pulse penetrate right into the hollow of the horn. The sensory system is involved too. The cow looks and listens inwardly, even with her eyes. She even smells herself, not the world around her. When she starts to eat, she breathes out, and smells back and then tastes what she has exhaled. So everything is directed to her inner organization.

We could think of the horn as being a type of sensory organ for the cow, where she meets the outside world to some extent, by pushing and feeling resistance. It is not really a perception of the outside world, but merely a pushing against something and thereby becoming aware. The horn contains a concentration of all the different systems active in the cow's organism. The horn is part of the head, the sensory system. The cow head is very occupied with the metabolic system with all that chewing and inner liveliness. If one observes dehorned cows, they are quite dull creatures. Through losing their horns they also lose their inner activity. Their forces stream out and are no longer structured by the form and substance of the horn.

The substance of the horn is almost entirely condensed protein, which is the fundamental substance of growth. But here the growth has died down to form this very dense substance. I do not

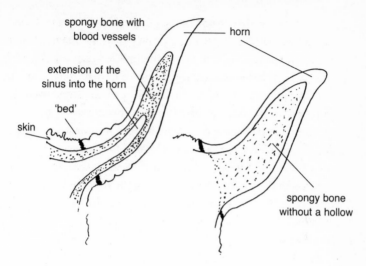

Diagram of a cow's horn and a bull's horn

think you will find such a dense organic substance anywhere else in nature. It is said that even radioactivity will not penetrate a cow horn.

Bull's horns

Let us compare a cow's horn to a bull's horn. Rudolf Steiner specified that a cow horn should be used, and not a bull horn. A cow horn is longer and thinner, more twisted and refined. The bone does not fully fill the space inside the horn, instead there is an airy space which extends almost to the top. A bull horn is wider, and much shorter and has a rougher texture, and the inner bone reaches right to the top without leaving a hollow space.

It is not only the form and structure of the horns that are very different, but also the behaviour of a bull and a cow. Originally, in the wild breeds, bulls had very large, heavy heads, and a very slim back. This is still the characteristic of male ruminants. They display a kind of nervousness throughout their life, and their astral body is filled with anxiety and fear. A bull, when he comes upon something which is new to him, is usually frightened, and this fearfulness causes the blood to stream from the rear into the head, and he makes an attack. This fear causes the bull to seek resistance in order to overcome the fear. He is far more exposed to the outside world than the cow. She rests herself and contemplates. We can observe a wonderful polarity between these two beasts.

It is quite obvious, if we regard the horn of the cow as an organ of intelligence, and at the same time related to the metabolic system, that it can only fulfil its function if the cow receives fodder produced on the farm. The cow reflects, in the form and substance of her horn, the living forces which are released by digestion and, indeed, the whole environment from where the fodder comes and in which she lives.

Manure

The horns are at the active pole of the cow, and at the other, something is released into the world absolutely passively. When you enter a cowshed the cows stare back at you dreamily, then all of a sudden you hear a 'splat, splat, splat.' This is the result of her perception, which stimulates the metabolic more than the sensory system. What she releases contains the result of what has been carried up by the blood stream into head and horn as a result of digestion. Whatever the cow has perceived within herself while taking apart the fodder by digestion impregnates the blood and is reflected by the horns into the metabolic system. This results in the manure, which becomes the carrier of this reflection. This process is just the opposite to human thinking! A human thought appears in our consciousness as the result of a dying life process. It is the projection of a life process into our consciousness. The inner wisdom of a life process is revealed in the content of thought. Therefore this content is dead. In the case of a cow, the life process she perceives stays down, because she has not got an individual ego. She can only gain consciousness of it on an astral level. She is not able to keep the result of her perception for herself, but releases it as manure.

Therefore the manure is a very special substance which, as we can read in the Agriculture Course, contains oxygen-carrying and nitrogen-carrying forces. Normally we think of oxygen as carrying the etheric and nitrogen as carrying the astral, but here it is the reverse. Her etheric and astral body are connected to nitrogen and oxygen. Thus the cow releases something which actually belongs to her, as a part of her own being. She does not keep it for herself, that is the privilege of a human being, but she sacrifices it to the world. The dung is a gift to the earth.

If one wants to work with the horn/manure preparations, one has to really sense these polarities, to feel the tension between them, and thus develop a personal relationship.

By the way, we should ask the question, why is the devil always depicted with horns? The devil can be looked at as being the opposite of a cow. Whatever he perceives in the world he is eager to keep for himself. What is rayed back into his own being by the horns is kept by him. When you look at a cow lying in the byres or grazing over the pasture, she emanates harmony and wisdom. The expression of the devil's faculty to keep everything for himself, this incarnated egotism, are the horns and hooves.

But why then has Moses been depicted with horns? Moses was a representative of the Egyptian culture, which was governed by the zodiacal sign of Taurus, the bull. But I don't think that is the real answer. Moses has a special position in the Egyptian culture as a representative of the Israelites, who were chosen to develop the faculty of thinking. They rayed back their life forces into themselves, but not to preserve them there, like the devil, rather to transform them into thoughts in order to communicate them on a new level. So intellectual thought can by symbolized by the horns of Moses.

We have discussed the polarity of the cow horn and the cow dung. Why do we fill the horns with dung in the autumn? Why not in the summer? Once again you need to be on your farm and experience its full reality. The cow herd is the inner life of the farm, as it were, because the cow feeds from the plants growing throughout the farm in the course of the year. This fodder is very different in

Burying cow horns in the autumn

quality as the year goes on. Normally the cows start to graze in the springtime and take in a very lush, protein-rich leafage. Later in the year the plants grow towards their seed and fruit formation. The cow participates in this process throughout the year, and responds to it. She does so not by looking at the grass and herbs, but by the perceiving activity of her mighty digestion process. This perception forms the 'thought' on the level of her astral being and from there it is released into the cow dung. In the autumn, nature ripens, and this ripeness is lifted onto this higher level and reflected in the cow's dung. It is a kind of fruit, a final thought of the seasons, because the cow has perceived etheric, astral and physical bodies of the farm organism, which have structured themselves in plant growth. This may explain why we gather the cow dung at this culminating time of the year.

We take something from the late summer, so to speak, and stuff it into the horn. We realize through our personal relationship that Michaelmas is the time to be making these preparations. A presence of mind gives you that inner certainty. I have experienced this as being a true way to celebrate Michaelmas, together with others whom you invite to participate in this process. One can have the feeling that this is an activity that belongs to everybody, not just the farmers. After filling the horns we bury them in the soil in the autumn.

Silica

Let us look at the horn/silica preparation. We have discussed the horn. It is obvious that the silica substance is quite different to the horn substance. Both are organic, but one is an animal substance, the other a mineral. They seem to be quite opposite, to represent a polarity. But now let us ask, where does silica occur in nature? It appears always at the surface of an organism, for example in the skin, eyes and hair and concentrated near the surfaces of plants, stems, leaves and in the case of grains, increasingly towards the periphery. All the sense organs contain a concentration of silica. Furthermore it is interesting to see where it appears in geological formation. Seventy-five per cent of all silica in nature occurs on the surface layer of the continents. What is its function? Rudolf Steiner pointed out in the Agriculture Course that the function of silica is to reflect the workings of the superior planets (Saturn, Jupiter and Mars), which support the sun. Silica benefits the growth of the plant, the development of the flower and fruit formation. It is a sense organ of the earth, as the horn is the sense organ of the ruminants, but the horn reflects and stimulates the inner activity of the animal, while the silica reflects the cosmos into the outer activity of the earth. So there is a connection in their function. Silica is representative of the past evolution of the earth, but reflects the present cosmos. The horn is a manifestation in the present, which reflects what the cow perceives while she is digesting the fodder grown in the past.

We collect some silica. Where do we find it? Beneath the earth, as crystals. Crystallizing forces work mainly in the wintertime.

Then it is the starry cosmos, not the planetary forces, that become related to the mineral world, they crystallize the water to hexagonal snow crystals, but also the minerals to crystals. There is a renewal of the earth every winter, when she is crystallized, right down into her depths. The quartz is the ideal manifestation in the mineral world of the crystallizing forces. This is a winter quality. The silica is a representative of the autumn/winter processes, whereas the manure is a representative of the summer/autumn processes.

The next step in making the silica preparation is to collect quartz around Easter time and grind it into a very fine powder. We reduce the crystalline structure to an amorphous state. It loses its capacity to reflect, in order to attain a new ability. We add water to it until it has a creamy consistency and pour it into the horn. We cover it with a layer of loam and bury it in the earth after Easter and leave it there throughout summer — not the winter — and expose it to the seasonal forces working there. Then we dig it up in autumn.

We have looked at the polarity between the manure and horn, and between the silica and horn. Now we should look at the polarity between the quartz and the manure, which is already indicated when we spray the preparations, the horn/manure on the earth and the horn/silica on the plants. Cow dung is initially amorphous, whereas quartz is an archetypal crystal. But looking at the function, we again find a certain relationship. The main constituent of cow dung is carbon, as well as nitrogen. Carbon is the carrier of formative forces. The main compound of silica is silicon, which has been the carrier of formative forces in the past, and is a remnant of the living sphere which has become a geological formation. Carbon is the carrier of formative forces in the present time. Silica and the carbon-bearing cow dung are closely related by their properties, but the one is concerned with the past and the other with the present.

We could say that silica is an expression of the contraction forces in nature, while the cow dung is an expression of expansion forces. One reflects the cosmos and relates to the fruit formation, colour and smell of the flower, the other is carrier of the forces that form the living structure.

Polarities

We have taken the first step in making the preparations and observed their ingredients and found polarities between them. When we grasp a polarity it can lead to a concentration of view towards intensification. This happens when we fill these substances into horns. We interfere in a natural process. Nature would never come to this idea to stuff the horns with these ingredients. It is not to be found in her wisdom that these two poles could be united in such a way. It is through spiritual insight, that we intervene in predetermined processes of nature. The next step is that the horns are exposed to the forces that are at work during winter and summertime. What happens to the dung-filled horn when it is buried? In wintertime crystallization forces take hold of the earth which becomes more physical and the ethers are released. This can be observed for instance when irrigating fruit trees on frosty days, because through crystallization warmth as an ether force is released. This freed ether rays into the cow horn and is concentrated in the cow dung and is thus preserved. The contrary happens in summertime when the chemical, light and warmth ether permeate the mineral world and thus the buried horn with its silica content also becomes permeated, preserving them instead of reflecting them.

Thus we create two new substances which cannot be found in nature. One is of summer origin and distributes winter forces. The other is of winter origin and distributes summer forces.

3. Stirring and Spraying the Preparations

In the previous chapter we looked at the first two steps of making the spray preparations. In the first stage we observed certain given polarities and how these are united through our actions. The second stage was to submit these substances to the polarity of summer and winter forces in nature. Thus we achieve the creation of new earthly substances. Subsequently we will take third and fourth steps along this path towards a deeper understanding.

Stirring the preparations

The third step, as indicated by Rudolf Steiner in the Agriculture Course, is to take this earthly matter in tiny amounts and add it to water. Then we start to stir the water rhythmically, mixing in the substance in such a way as to build up a vortex, creating a funnel in the centre. Finally we have the liquid ready to be sprayed out. We begin to stir when the appropriate time to apply the preparations has arrived. This depends on the sowing time and the growth stages. By stirring, the earthly substance is transformed into a liquid state. The first step in making the preparations was of our own doing, the second step is given by the seasons. This third step requires our complete commitment. It is especially during the stirring of the preparations that we can build up a personal relationship which leads us to a deeper understanding of what we are doing. From this point of view it is the most crucial stage of the whole preparation-making.

These are the tools we need: a vessel, better made of wood or of food grade plastic rather than metal. We take a broom handle and fix some birch branches to the bottom, and attach the broom handle to the ceiling or a rafter. We fill the vessel with water, preferably rain water. If there is none available, or if it is too polluted, we could take pond or well water. I consider it to be ideal to activate the water before stirring it using a Flowform cascade.[1]

We start to stir the water by building up a vortex. Then we break it down and stir in the other direction. Doing this we are creating a polarity in space and time. The polarity is formed in space by creating a centre and a periphery. In time it is formed between the water in stillness and in motion, forming the vortex. Another polarity is created by making a clockwise vortex, and then an anti-clockwise vortex. A further polarity in time is the speed of the water, moving very fast at the centre and much slower at the periphery, thus forming gliding layers at different speeds. So we are actively creating polarities in space and time, tensions between two poles out of which something new evolves. This tension is a medium, so to speak, through which an intensification can occur. What kind of intensification are we aiming at while stirring?

We must try to imagine that an inner structure of infinite planes comes about in the water. It is no longer merely a body of water, but a plane in space and a tension in time. The plane in space is created by an expansion in time, and the water loses its relation to the earth and is lifted towards the airy element. Water has its centre of gravity within itself, but the air does not. By increasing the speed of the vortex while stirring we create an inner sensitivity on the surface between the liquid and airy state. We build up and break down the vortex again and again. According to my understanding where this sensitivity occurs two realities meet. The earthly matter of the preparations that we have created, horn/silica (501) and horn/manure (500), meet another reality to which the water has become sensitive. These are the forces acting at the present moment. Before stirring, the preparations that were made during the previous year, were kept in a vessel. Once stirred, they have to be applied within a certain time. So while stirring, the earthly matter of the preparations is meeting the present cosmic environment. We transmute the earthly substance into a liquid state and by so doing we are opening the water to the present workings of the cosmic environment. If we leave this process to a machine we give up the opportunity to experience this. We are a part of this process of building up a sensitivity arising out of polarities. In this a trinity meets: the earthly realm, represented by the substances of the preparations; the cosmic realm with the whole cosmic periphery at the very moment when we stir; and thirdly your very own self that is willingly involved in this process.

The activity of stirring

As with the stirring we also create polarities within ourselves, in the inner activities of thinking and will. When we take hold of the broom handle and start to stir, moving perhaps 200 or 300 litres of water, we take a conscious decision to move the water. We push the thinking into the will, and the will is pushed by our consciousness. We externalize our will in the vessel. In building up the vortex it is necessary to move the water faster, the will has to be in advance of the moving water. We pull the water behind us and accelerate it according to our personal rhythm. When we reach the limit of our will to build up the vortex, we break it down. If you observe yourself you will see that your consciousness is not completely involved in reaching this limit and breaking down the vortex in the same way as it is involved in starting to build up the vortex. The faster you stir, the more your conscious thinking disconnects from your will, and becomes still within you. The will becomes externalized through the physical activity. It is an energetic job and you start to sweat. On the other hand your consciousness internalizes and you are now still, and a supreme observer of your will activity. We feel a tremendous tension of this polarity within us which corresponds to the building up of the vortex. Once you break down the vortex this inner tension is also destroyed.

Out of this tension arises an intensification towards a higher presence of mind. There, in that tension, our feeling expands between internalized thinking and externalized will, while our feeling is an expression of a higher objectivity and is revealed as a presence of mind. Observing yourself you will realize that you cannot feel a feeling from the past, nor can you feel the future as future. You can only feel the present. Thinking relates to the past. Will is directed to the future. The feeling is the element of the present. In stirring by hand we create a new quality of objective feeling, which is derived from this tension. So for instance, while stirring, you can have very pertinent ideas, all of a sudden. You may perhaps have an impulse to look out on the pasture at what the cows are doing or you may look at some remote part of the farm and, doing so, it proves to be right. Through this presence of mind we feel 'inside' the reality of the farm. It is a very special experience, but quite impossible to maintain for a whole hour of stirring. The more often you stir, the more you can expect to attain and maintain it.

One problem is that sometimes people tend to do the stirring in a very serious mood, like a religious rite. Of course it is a virtue to be earnestly involved, but this can easily lead to a subtle dogmatism. It is far more appropriate not to stir on your own, as you better achieve a presence of mind and the right mood if you stir together with others. That is why, when we started our farm back in 1968, we decided from the outset always to stir with at least three people. We said we do not mind mechanizing whatever farm operations we can, but never the stirring. We organized the farm in such a way that we always had the time for three people to stir together. It is of great importance, within a mechanized farm, to create a free space in which to become creative. The work with the preparations affords us this free space. As the work with the preparations is a matter of free will, we must learn to establish this free space freely. Then the stirring becomes a social process, a melody starts to lift you up and make you a little lighter, be conversing, telling stories and developing a mood of cheerfulness. Thus we work towards building up a community while stirring. Then the spiritual atmosphere comes about which is conducive to creating the presence of mind.

The stirring continues for an hour. It is strenuous work, especially on the first occasion in the year in springtime, when you are no longer used to it. The second and third times are a little easier, and in the end it almost happens by itself. After some time, you gain a different relationship to the whole process. You can have the feeling that it is not something trivial that we have done, but something that is really a building stone to construct the farm entity. In doing so we work into the reality of the farm individuality, through a sacrifice on our part.

Stirring the preparations is not something that is dictated to us by a natural law. You yourself out of spiritual insight determine when and what to do. And that is working towards outer freedom. Whatever else you do in agriculture necessarily follows some natural law. The calf tells you it needs to be fed, the field and weather tells you when to plough. The work is determined by natural laws and the more you follow them the more successful you are. Yet you are unfree. But this is not the case with the preparations, where you yourself determine what to do. This is creative in a true sense and we embark upon a process towards freedom. Thus it may become a celebrations in which others should join. Not only your fellow co-workers on the farm, but invite other people from outside. We have an 'agricultural community' around us, and they often participate in this social process of stirring. It should become a custom that people from the cities come and do the stirring together with farmers. It is an experience that cannot be gained anywhere else. It opens a door to nature, and people can really exercise their responsibility for the earth. So we continually create festive hours amidst the heavy toil of the normal working week.

Spraying

By stirring for one hour we transform the solid matter into a liquid state by expansion and contraction. The next step is to transmute this liquid into an airy state. The liquid is spread out into droplets, which stay in the air for a short time. The length of time is immaterial. The droplets mirror the whole periphery at this very moment. They mirror the quality of light, warmth and air according to the time of day and conditions of the natural environment. A transition from the liquid to the airy state occurs by spraying. The droplets then fall onto the soil (500), or the plant (501). While stirring you are intensively involved in the process and you have all the liquid in front of you. When you spray it, it disappears. Wherever the droplets land, they are sucked in by the earth or evaporate, while the essence enters into the warmth element.

What have we done? We have taken earthly substance, transformed it into liquid, transmuted it into an airy state, and then transferred it to the warmth at the very spot where the droplets landed. We are reversing the whole path of the Earth's evolution. We start off at the end of the earthly stage. We prepare earthly substance which has become disconnected from its spiritual origin, and we guide it back. We cannot think of the droplets as migrating down into the soil and somehow, by cause and effect, reaching the roots and being taken up the plants that are growing there. To understand it as a physical chemical process is impossible. One has to be aware of the fact that we really lead this substance through the gateway of warmth into the etheric realm, to the border of space and time. All the way along we have been following processes of expansion and contraction. We expanded the liquid into the atmosphere when spraying, and then it was contracted in the warmth of the spots where it landed. There again it expands into the etheric world. What happens in this, to our physical senses, 'unknown landscape'?

The elemental beings

To find an answer, let consider the plant, forming its leaves and flowers, its shape, scent and colours. It unfolds when earth and heaven meet. When seed formation takes place and the plant dies it fades to the ground, and all the events, the cosmic ideas that become materialized in forming and shaping vanish. In the colour and forms of the plant we perceive these cosmic ideas as an outer appearance and produce a thought which originates in the same realm as these cosmic ideas. The latter are hidden in the outer manifestations of nature. The life processes out of which a plant grows are also within you. When they die down into your consciousness they reveal their inner being. When the plant dies down and mineralizes or decays into humus, these forming and shaping cosmic ideas are released into the etheric world, where there are beings, servants, as it were, at work.

In former times humankind gave names to these servants, or elemental beings, who build up the plant out of the cosmic ideas. Their task is to preserve the cosmic ideas throughout the winter. They are called gnomes, undines, sylphs and salamanders. Humanity knew about these beings in earlier times when they were not so scientifically developed as we are. Although they were not as sophisticated as us they had an inherent wisdom, because they still had a natural relationship to the elemental world. Rudolf Steiner gave some lectures about the significance and workings of these beings in 1923.[2] Incidentally, I think that most of the people who subsequently attended the Agriculture Course in Koberwitz thought that Rudolf Steiner was going to continue with these talks, describing how the farmer, because of his familiarity with these beings, knows how to work with them. But Rudolf Steiner did not say a word about the elemental beings, except once when speaking about the yarrow. Then he touched on the subject by speaking of the nature spirits, which was the only direct reference to them.

Rudolf Steiner describes the elemental beings, and the gnomes especially, as having no morality. They depend on the morality of the hierarchies and of human beings. When we expose our will in work we affect their realm. So the elemental beings depend on what kind of morality they meet, and can be servants of either good or evil. We are only able to fulfil our responsibility to the earth and the environment if we know about these elemental beings, and deliver moral impulses to them. Imagine, therefore, how vital it is to really build up a personal relationship to nature.

The droplets of our preparation are absorbed into the etheric realm. They are a fertilizer, so to speak, for the elemental beings. It gives them new spiritual guidance for their work. So they are once again able to concentrate their work where we want them to help: in plant growth, in fruit formation and so on. So we start to direct this unknown world, the elemental beings, by 'manuring' this etheric realm. In a similar sense we can see the working of silica. Preparation 500 is sprayed onto the 'head' of the agricultural individuality, and works in the solid and liquid elements in the world of darkness, underneath the earth.[3] Preparation 501 works into the atmosphere in air, light and warmth, up in the 'metabolism' of the agricultural individuality. Rudolf Steiner describes silica as the 'outer sense' of the earth. When we transform it into preparation 501, which is a receptor of forces that work in summer, we then 'manure' the 'belly' of the farm with these forces. We organize and direct the elemental beings at work in this sphere, especially the sylphs. I once found a description by Rudolf Steiner of the workings of the elemental beings. He said the gnomes push upwards, and the sylphs draw out of the atmosphere ... this is exactly the same as the words he used to describe the working of these two preparations in the Agriculture Course. I am quite convinced that this is the main activity of these preparations, to guide the servants of nature working side by side with the spiritual hierarchies.

The right time for spraying

I would like to conclude by asking, what is the right time to apply these preparations? Don't expect a recipe. We are always so keen to have recipes for doing these things. The right time for application is a question of your individual relationship to your farm. Of course one can say that the horn/silica should actually be sprayed in the morning, with the rising sun, when the dew is falling and plant sap rising. It is quite obvious that the preparation 500 should be sprayed some time in the afternoon, but there are many possible variations. Everyone can have their own opinion and picture of how to work with them. On our farm, I have experienced that when I have worked throughout the day, and I walk across the fields in the evening, looking back on the work of the day in a contemplating, pensive mood, then the intuition could come that tomorrow is the day to spray the preparations. You have to be fully involved in the whole farm, so that the farm organism is your extended body. And you have to create a consciousness of this extension, and then you will have the presence of mind to know when it is the right time to spray. Then maybe you stir at some time during the day and you feel free in this decision because you yourself determine it out of your own insight.

We have tried to take some steps on the path of making the preparations, and looked at the landscape to the left and right. We found that there are given polarities, for instance horn and cow dung, winter and summer. Then we continue on the path, and the further we go the more we ourselves are committed. We ourselves create some of the polarities and build up our own landscape. This personal relationship forms the ground of our understanding. It can give us an inner certainty, which creates a new identification between us and the reality of the farm. That is what we must strive to achieve, to overcome this abyss between us and the outside world. Much of our work tends to widen this abyss: when working with machinery, we are on one side and the reality is on the other. We bridge the abyss with modern technology, without knowledge of what we are doing. Only afterwards nature herself tells us what damage we have done to her.

But here, in working with the preparations and building up a personal relationship, we follow a path and, step by step, cross a bridge into reality, and plant a germ there in the present for the future. This is the deeper meaning of manuring. The earth has come to its end and has no future out of itself. So it is a matter of freedom to sow the seed, the germ, to give it a future so it can develop together with us. It is due to this aspect of future evolution that biodynamic farming becomes the most creative work we can think of. Although we are right at the very beginning, the perspective of biodynamic farming reaches far beyond what most people think, in our modern times, to be progressive. In biodynamic farming we can have the inner certainty that the deeper we enter into spiritual insight, the more we are serving an evolutionary process. Once you have grasped that fact, why not become a farmer?

4. Life Forces and the Land

The story of the man who planted trees is strongly connected to biodynamics, because the way Monsieur Boufier just did what he thought was right is really a blessing. It is a blessing for modern man when one has such a certainty, such an inner knowledge, not to be too far away from the truth if one just does what one thinks. This virtue to do and think, and to think and do requires courage, and that is greatly lacking nowadays. In biodynamic farming, and especially our work with the preparations, we have to evoke the same sense of certainty. With the preparations we are doing something without seeing the immediate result. We take tiny amounts of earthly substance which look like humus and we put them into holes in the compost heap. We cover up the holes and go away, accompanying the natural process which now goes on beyond our perception with a feeling of certainty that something most significant will happen. Our whole modern civilization is a desert, and will remain so if here and there we do not insert tiny little germs of ideas and deeds, while being deeply convinced that they will come out one day and flourish.

Forces of decay

We have read in the Agriculture Course that the preparations work with forces. Forces are not visible. We only perceive their effect. Sunlight as a force is invisible. What we see are colours that appear when the forces of darkness (physical matter) and light meet. In order to understand what forces are, we have to behold ourselves. In carrying out an action we reveal our will. We experience our will as being the force that creates a certain effect. At the same time we are sure that this force has its origin in our spiritual being. In observing ourselves we recognize that the effect of the forces we show are of spiritual origin. In the outer world we see the effect while the inner nature of the force and the spiritual being creating this force are hidden.

How are we able to identify different qualities of forces in the outer world? For example we can observe a dog. Let us first imagine a dead dog. When so doing all the effects that appear are decaying processes: forces of disintegration appear. They end in complete mineralization. The ruling laws of this realm are cause and effect. The forces at work are mechanical ones, leading into sub-nature: the forces of electricity, magnetism and radioactivity. The effects of these forces of decay are finite and therefore calculable.

If we merely make abstract theories of what we experience in ourselves and what is hidden in the outside world, we create a new realm below nature. This is the mechanistic realm of technology. This realm benefits the social life but in the long run it brings about destruction and disintegration in our natural environment.

With the theoretical ideas gained from the decaying processes we cannot construct a living dog, only a machine. We are able to isolate and even handle the forces that are active in the mineral realm. Indeed we are concerned in our daily lives with these forces, namely electricity, magnetism and nuclear power, and of course the mechanical forces, although nobody understands their inner nature. We have to ask, what beings are creating forces which ultimately lead to destruction or mineralization?

Forces of life

Let us now take the other case and observe a living dog. It reveals forces out of an inner activity. We see it breathing, regulating its own warmth organism, and how it feeds. Furthermore, there are life processes active in the dog's organism; those of inner secretion and excretion, and of life maintenance: growth and reproduction. Observing the dog we can see that these processes keep the organism alive. Through them the organism appears as a unit, a wholeness. Whatever life processes are active are embedded in rhythms. The forces at work in the decay of a dead dog will not manifest rhythms. The rhythms are congealed into a beat and are therefore calculable. Life processes do not follow the laws of cause and effect, but rather the law of simultaneity. Take, for example, the physiological processes in a cell of the liver, which is the most active organ in the organism. The size of a liver cell is about one-hundredth of a pin-head — simply a tiny dot. Within such a cell, about twenty different physiological processes have been observed all happening simultaneously. There are those that build up glyco-gen and those that break it down, both acting at the same time! It is impossible to think that this all actually happens within that cell. The liver cell is just the physical point, and all that happens has its origin in the cosmic circumference far beyond. In order to under-stand the physiological processes in the liver cell or any life process we must seek its origin in the cosmic circumference.

Observing a dead dog, we do not see the decaying forces as such. What we perceive is a sequence of images that we produce out of an inner activity. By perceiving we grasp the appearance of a decay-ing process, but the forces that are at work are hidden. The same is true when observing the living being. When the dog is barking or jumping at us, again we develop through inner activity pictures which refer to the appearance yet not the inner, ensouled life of the dog as such, the astral body. What we become conscious of by thinking touches the surface of a hidden world. Our thoughts are just as dead as the perceived objects. But the crucial point is that, while we are building up images, the same force that is active in thinking, is active in the outside world as the force of plant growth.

In thinking a life process is active. The thinking activity is an inner life process which remains unconscious. The very moment it comes to an end, that is when it becomes conscious, it dies into the thought. The inner wisdom of this life force is revealed as the con-tent of the thought. From this point of view life forces are active wisdom. The forces in thinking are related to the forces that allow the plants to grow. In plant life they also die, but into blossom, whereas in thinking they die into the thought. In the flowering stage the wisdom of the plant-being is manifest. In the same sense as plant growth ends in the blossom, so the thinking activity culminates in the 'blossom' of thought. Both are revelations of reality on two different levels. That is why our thinking can lead us to the truth in nature. The force active in thinking dies into the thought, while when feeling we are in a more dreaming state and the forces reveal more of their inner quality.

Ethers of a landscape

Let us now take another example and imagine going out and climbing a hill and looking into the surrounding unspoilt landscape. We immediately perceive it as a wholeness. We do not see isolated segments but an inner coherence. The force creating this coherent image of a wholeness may be called light ether. Outside the landscape is flooded with light. At the same time, looking at this landscape it is the hidden force of the light ether that builds up all the different manifestations into one wholeness. Just as the light connects everything in the outside world, it is the light ether within me that builds up the image of this wholeness.

If we continue to look at the landscape throughout the seasons we have a different view in spring, summer, autumn and winter. This points to a transformation taking place, which reveals the working of chemical ether, which is another quality of force in the etheric, in the life realm. The forces that cause metamorphosis in the outside world are active within us when we can see this metamorphosis. If we look at the same landscape from another point of view we see that every single feature is related to the others. For example, there is a meadow and here is a field and beyond is a valley and a distant hillside covered by a forest. This immediate feeling that everything is interrelated is a reflection of a force of sense activity in the outside world. It is the revelation of the life ether which is active in us also when we recognize the relationship between things.

If we look at the outside world and it appears to us as a harmonious wholeness, the ether forces are in a healthy relationship. If this coherence is disturbed then decaying forces predominate and our living realm falls ill. Decaying forces belong to the mineral world. It can happen all of a sudden that the inner coherence of a landscape breaks down. The dying forests are such phenomena. The forces of light, chemical, life and warmth ether are no longer able to maintain this inner coherence of the landscape. Something similar happens when we cultivate the soil. We activate the oxygen process and thereby break down the stable humus. Forces are being lost along with substance.

light ether	—	landscape
chemical ether	—	relationship of processes
manifestation of ether	—	harmony of landscape
when this breaks down	—	dying forests

Fertilizer

The task of fertilizing is to restore not only substances, but also forces to the soil. Mineral fertilizer mainly restores substances and physical forces which, in the case of salts, promote decay. Take the example of artificial nitrogen. It immediately dissolves water in the soil and thereby influences the electro-magnetic potential in the soil. This undoubtedly affects the formation and sensitivity of the plant root hairs and its ability to perform symbiosis. If we apply potassium salts, an increase of radioactivity can be detected because of its natural radioactivity. Thus mineral fertilizer, while supplementing diminished levels of minerals, does not supply forces of any value. Organic matter from the living realm also replaces living forces. Ensouled matter from the astral realm, for instance cow dung, introduces living and ensouled forces to the soil. That is why cow dung is so indispensable in biodynamic farming. Soil fertility is transformed into plant growth. The question is, how can we restore these forces which were composing the substances in the stable, crumbly humus? And what about those substances and primarily forces that have really been lost because produce has been taken away from the farm? How do we restore these living and astral forces? This question is very connected to another: how can we improve the soil in such a way that the earthly substance becomes susceptible to the living forces which have their origin in the cosmic periphery? Or how are we able to make the plant sentient so that it can actively find the substance it needs? And above all, how are we able to enliven the earth so that the dead mineral matter is vitalized?

These questions are not a matter of restoring something that has been there and has got lost. They demand another answer. We have to develop a technique to make physical substances susceptible to living forces. This technique was well known in olden times. A couple of centuries ago in true alchemy the alchemists were seeking to vitalize substances to become remedies. One of the most outstanding representatives of these techniques was the Swiss physician Paracelsus (1493–1541). A follower of his, and a contemporary of Goethe's, was Samuel Hahnemann (1755–1843) who developed homeopathy on an empirical basis. He carried out most of this experiments on himself, for instance he poisoned himself with arsenic then healed himself with highly diluted arsenic. So he formulated the healing law of homeopathy, *similar similibus,* which means 'like cures like.' This empirical homeopathic approach has been put on a sound scientific basis by Rudolf Steiner. His spiritual research was the foundation of anthroposophical medicine and following his indications many experiments were made by Lili Kolisko.* The procedure followed to prepare homeopathic remedies are as follows.

* Lili Kolisko (1889–1976), was the originator of capillary dynamolysis, a scientific technique, which has been applied to medicine, agriculture and other spheres of science. She and her husband, Dr Eugen Kolisko (1893–1939) worked together with Rudolf Steiner at the end of the First World War.

When we want to make a remedy out of certain substances, for instance iron or silica or some extract of a plant, we take one part of the particular substance and dilute it with nine parts of water. (We can take other mediums for dilution instead). Then we shake it and thus obtain the first decimal potency, known in its abbreviated form as D1 (or X1). Then we take one part of the D1 solution and dilute it again with nine parts of water, shake it well; this results in a second stage of dilution, D2 (or X2). So we can continue to ever higher levels of dilution. For instance if we reach D18 it is the same as if we take one gram of substance and dilute it in the Lake of Constance. If we come to about D21 we enter the realm of the Loschmidt constant which indicates that no molecule of the original substance remains. And yet we have specific effects. Forces continue to be active although none of the underlying substance can be detected any longer.

Homeopathy has become the basis of anthroposophical medicine. Out of spiritual research, Rudolf Steiner indicated that different stages of potencies have different healing effects. So quite low potencies, around D7, are effective for metabolic illnesses. For disorders of the rhythmic system, such as a heart or lung malfunction, one would take potencies in the region of D12 to D18. The very high potencies heal illnesses or inflammations of the sensory system. So one and the same substance may develop different healing properties according to how highly it is potentized.

metabolic	rhythmic	sensory system
D7	D12–D18	D30

What is the healing process for human beings? It is when your soul-spirit is able to incarnate fully into your body and becomes a supreme governor of all bodily processes, and of how the members of the body relate to one another. As soon as something is in disorder, most usually caused by an imbalance between the astral body and the etheric and physical bodies, then you become ill. The soul-spirit is not strong enough to build up the substances to become carrier of the ego-forces. Therefore you have to introduce remedies from the outside world and prepare them in such a way, as we have described, that the ego forces are able to promote the healing forces.

This is the case with human medicine. But what is the medicine for nature? How can we heal a deficiency of forces in nature, in all that we are dealing with as farmers: the animals, plants, the soil and the farm organism as a whole. Where is the being active for which we are producing healing substances? Looking at the human being, we know that the ego forces are able to individualize any substantial processes. For example each of us has our individually structured protein. No two people are the same in this respect. The ego forces are structuring and individualizing all physiological processes, and the composition of substances and forces, right down to the last cell. All is composed into an ego-organization. But in nature nothing is in this same sense individualized. So where is the being that is able to individualize, so to speak, the life forces and substances in the outside world? This is the starting point of the question of why we use the preparations.

Remedies for the land

As in medicine we produce remedies for our own sakes, we must as human beings prepare and individualize substances for the sake of the outer world, which then become susceptible for forces which are nothing other than the deeds of spiritual beings. So we have to search for and prepare substances which can become the physical carrier of soul-spiritual forces in the same sense as our individual bodily substances are carrier of our spiritual being, our ego.

Nature provides all the material, and takes the first step towards such an individualization, but is not able of fulfil it. We have to add the spiritual concept. The materials that nature provides are certain mineral substances, for instance silica which we use in the spray preparation 501, or the herbs for the different compost preparations out of the plant kingdom, and animal organs from the animal kingdom. Nature also provides the rhythms of the seasons, the elements above the soil of warmth and air, and the elements below the soil of water and earth. So all the constituents to make the preparations may be found in the outside world, but the spiritual concept of how these constituents are joined together is not hidden in nature as a natural law. It has to be found by spiritual research beyond nature in that realm where the being of man originates.

The first step is provided by nature itself. Different plant species have abilities to work with different earthly substances. This ability is the basis for the healing properties in our medicinal herbs. They are able to potentize earthly substances quite naturally within their living processes from the roots right up to the blossoms. The yarrow is able to work with potassium together with sulphur in a unique way. It is able to potentize the potassium, by its own living processes, to an ever higher stage, but then it comes to an end in the blossom. The camomile is able to potentize potassium and calcium; the stinging nettle, in addition to potassium and calcium also works with iron. it is specialized to potentize the iron process in nature and is therefore a kind of remedy for an abundance of iron radiation. The oak is able to work with calcium in a special way, as it is found in large, well structured quantities in oak bark. The dandelion governs the potassium in relation to silica; valerian masters the phosphorus process. These special abilities of the different plant species are revealed through spiritual research; their own life processes are able to enliven within themselves, and potentize to a higher level, the dead, earthly substances. But this can only be achieved up to a certain limit, the flowering stage, and then it ceases.

Forces of the future

There are another two aspects I would like to refer to. A spiritual one which I will look at in more detail in the next chapter, and a practical one.

The spiritual aspect we find in Rudolf Steiner's letter, 'What is the Earth in Reality in the Macrocosm?'[1] He wrote these letters shortly before his death to members of the Anthroposophical Society, about the significance of the plant, mineral and animal kingdoms in relation to the formation of a future macrocosm after the development of the earth has come to an end. There he wrote the following. The plant grows from seed to seed. It germinates from a seed and ends up as a seed. But all the germinating forces that are at work during the growing process do not concentrate completely in the future seed; there is a superfluity which is not used up, so to speak. These excess germinating forces are donated to the etheric world and form the substantial image of the future macrocosm. Then he speaks about the mineral kingdom and says that minerals also ray out germinating forces as a surplus, and they direct these substances into a framework of the future macrocosm. The animals, especially the warm-blooded animal kingdom, do not emanate a surplus of germinating forces, but donate forces of form that envelop and create a sphere around this future macrocosm. So the earth is not only the end of something, but at the same time it is as a whole, a germ for the future. Are we able to retain and work with these germinating forces in making the preparations? Is the inner secret of making the preparations, that we are actually harnessing these germinating forces right now, so to make use of them in our manuring? Is it these germinating forces, which normally ray out and are preserved for the future in the etheric world, that we are able to introduce into nature via the preparations as a means of evolution? That is the question I just wanted to pose now, and will return to in the next chapter.

Results of preparations

I will now refer to some results of the working of the preparations. Since the beginning in 1924, experiments have been carried out on the effects of the preparations. Of course, above all a personal relationship has to be built up by anyone using the preparations: it is like practicing an art. But besides practical experience, scientific experimentation has proved to be of great significance. It has helped to deepen our understanding, to make applications more accurately and promote a sound public discussion. Especially after the Second World War, quite extensive experiments were carried out on the preparations, in connection to other questions. I am sure some of you will have had some observable results while working with the preparations. For instance, that within a very short time the manure loses its pungent smell, that the whole microbial decomposition processes are working in a different, more harmonious way. Bo Pettersson ran a series of field trials with crop rotations, together with the Agricultural University of Uppsala in Sweden. This experiment ran for thirty-two years, with different plots treated in different ways: one was fertilized with mineral manure, one with organic manure, another with organic manure and the biodynamic preparations. The result of this long-term experiment is quite striking. I mentioned earlier about images and the fact that we cannot see the forces themselves. but we can see the result of these forces. Bo Pettersson was able to show, after twenty years, that not only the humus content increased in the biodynamic section of the trial, but the fertile layer of the soil profile has deepened considerably. The humus content was not only concentrated at the surface, as in the minerally-fertilized plots, but reached much deeper down. This points to the fact that, as an effect of the compost preparations, the roots go far deeper in the soil, are far more finely distributed and make use therefore of a much larger soil volume.[2]

Another experiment started in 1980 at the Institute for Biodynamic Research in Darmstadt, Germany. Initially the main question was to find out the effect of compost preparations on food quality. This experiment was supported by the government, and carried out using the highest standards of modern scientific methodology, so it is fully accepted in academic circles. It was planned to run for four years but was extended to eight. The results showed considerable influences on the physiological fruit formation. When they had finished the experiment, the land was left uncultivated and at the beginning of the next season they went and looked at the area again, just by chance, and saw that the biodynamic plots in the statistical arrangement had a darker colour than the mineral and the organic plots.

They wondered how is it possible that after eight years on a sandy soil, the biodynamic plots which were scattered all over the area, showed a darker colour. So they began to experiment again, on the humus content at different depths and its quality on the root formation and on the microbial processes that take place. It is still going on, but the first result shows that the humus content after eight years remained the same in the biodynamic sections as it was under normal biodynamic conditions at the beginning of the experiment. In contrast to it the humus content of the organic plots dropped considerably and even more in the mineral plots. In the organic plots the same amount of compost was applied, with the same nitrogen level as on the biodynamic plots, but without the compost preparations. The organic sections had a higher humus content than the mineral plots, where the nitrogen had been added, but in different forms. Merely by applying the preparations to the compost used in the BD plots over eight years, the humus content had comparatively increased, not only in the top soil but also in the subsoil. The same result was observed as in Sweden, that the roots really penetrate into the soil.

The second result was that the quantity of roots was more or less the same in all three sections, but in the biodynamic section the amount of fine roots, root hairs, etc. was double that of the mineral section. It shows clearly that the roots are far more spread over the soil and subsoil and are thus far more intensively in touch with the earthly realm.

The third result is the most extraordinary. The soil respiration, that is the release of carbon dioxide produced by the roots and microbes, was the highest in the biodynamic section. This normally indicates an immense decomposition of organic matter. But here there was a greatly increased respiration at the same time maintaining a higher humus content! That means that the microbial life was not doing what it normally does, decomposing. It behaved in the opposite way, ruled by a higher force to which it became servant. This is what Rudolf Steiner describes in the Agriculture Course about the effect of the compost preparations, mainly the first three, yarrow, camomile and stinging nettle. He says that these three generate *nous*. The soil becomes sensible. So the microbes do not act one-sidedly according to their decomposing ability but respond to a higher order. This higher order releases forces that direct the microbial life to build up stable humus and to sustain soil fertility on a high level. These experiments, which were set up on a sound scientific basis, have demonstrated what has been achieved through spiritual research: the compost preparations as a whole work in such a way that the physical substance, the manure and finally the soil, become susceptible to forces of etheric and astral origin that are normally not, or not to such an extent, active in the soil.[3]

5. Protein and Yarrow

I would like to use the yarrow preparation as an example to show some of the principle of the preparations, of their makeup and usage. Rudolf Steiner lays the foundation for our understanding of the yarrow preparation, and indeed all the other preparations, in the third lecture of the Agriculture Course. This is when he speaks about the archetypal substance of life altogether: protein. The life of the animal and human kingdoms would not be possible without this archetypal substance in the plant kingdom. As human beings we need to take it as a constituent of our food. We break it down completely in our intestinal digestion. This active breakdown of well structured protein from outside enables us to build up our own protein. It is not the substance of the protein that is of any value, but its pattern. Its specific structure delivers a kind of model, which we perceive especially with the liver while the protein is broken down. In fact there are four organ systems that are involved in perceiving the protein pattern and they are able, in so doing, to build up our individual human protein. These are the kidney, the lung, the liver and the heart systems. All four are able to build up substance as it were out of nothing, yet with the basis of the form pattern, this highly individualized protein. This is the reason why this archetypal creation has to take place first in the outside world, in the plant kingdom. Four elements correspond to these four organ systems within us: carbon (kidney), oxygen (lung), nitrogen (liver) and hydrogen (heart). These four elements, together with sulphur, constitute protein. In the Agriculture Course Rudolf Steiner calls them four (or five) sisters.

They are also broadly distributed in the outer world. We breathe nitrogen (79% of the air is nitrogen); we breathe oxygen; we find carbon in coal, sulphur in an elemental state, and hydrogen is everywhere. All these elements are carriers of forces. Carbon is the carrier of formative forces, oxygen of life forces, nitrogen of astral of sentient forces, hydrogen is close to the physical and relates at the same time to the spiritual world. It is the carrier of the forces that are released from the physical world into the cosmos. In contrast to hydrogen, sulphur is the mediator of the spiritual with the earthly realm. In the outer world these elements are more or less separate from one another. So it is a mystery how to combine especially the four elements, carbon, oxygen, hydrogen and nitrogen. These are the basic constituents of protein. In the plant kingdom this process involves two substances that are polar opposites: sulphur and potassium. Thus we see that protein is formed when the three principles that were once well known in alchemy are equally at work: the salt principle — potassium and related elements, the sulphur principle, and in between the mediating mercury principle. Sulphur and potassium, as representatives and carriers of cosmic and earthly influences, help the mercurial archetype, the proteins, to come into being.

While speaking about protein Rudolf Steiner introduces the compost preparations and refers to the mystery of how protein comes about in the right way in the outside world. He says there is a most miraculous plant which can be seen as a model for all the other plants, in the way it handles the sulphur and potassium process with regard to the formation of protein. He characterizes this miraculous plant, the yarrow (*Achillea millefolium*), from the point of view of spiritual research and I would like to read his description.[1] He says:

> Take yarrow — a plant which is generally obtainable. If there is none in the district, you can use the dried herb just as well. Yarrow is indeed a miraculous creation. No doubt every plant is so; but if you afterwards look at any other plant, you will take it to heart all the more, what a marvel this yarrow is. It contains that of which I told you that the spirit always moistens its fingers therewith when it wants to

carry the different constituents — as carbon, nitrogen, etc. — to their several organic places. Yarrow stands out in nature as though some creator of the plant world had had it before him as a model, to show him how to bring the sulphur into a right relation to the remaining substances of the plant.

One would fain say, 'In no other plant do the nature spirits attain such perfection in the use of sulphur as they do in yarrow.' [It is very interesting that Rudolf Steiner mentions the nature spirits at this point, and it is the only place in the Agriculture Course where he does so.] And if you also know of the working of yarrow in the animal or human organism — if you know how well it can make good all that is due to weaknesses of the astral body (provided it is rightly carried into the biological sphere) — then you will trace it still farther, in its yarrow-nature, throughout the entire process of plant growth.

The appearance of yarrow

Anthroposophy widens our view and so first of all we should look at whether the yarrow reveals some of its miraculous nature in its outer form. When looking for yarrow, we see that its natural environment follows human civilization. It appears in the open plains where sun and earth meet directly, in meadows which are only cut once — not so much in permanently grazed pastures — along the roadsides and field borders; in an open area where sun and earth meet directly. We find yarrow more in dry that wet areas, more in loamy that sandy soils, and more in sunny that shady places. When we observe the plant itself, we see that the yarrow germinates in spring time and first forms a rosette. The leaves are pressed flat to the ground. All of a sudden, usually around the middle of June — this depends on the region of course — it shoots up very quickly and forms the first flower buds. Normally the yarrow blossoms during June, July and August, when the earth-cosmos relationship is at its greatest. But we can often see the yarrow flowering in September, October and even into November. When it wilts in late autumn or winter, what is left? We see the ligneous, solidified stems gathered in clumps all over the meadows which give the image of very sturdy plants.

Looking at the leaves, a real metamorphosis is visible. Although the leaves split up from a midrib and almost disperse towards the periphery into a range of leaf stems that fan out into tiny little spears, they show a complete leaf blade. The manifold pinnation ending up in points and spears very much relates to a strong sulphur activity. If we observe these points and spears in greater detail with a magnifying glass we find astonishingly an almost succulent shape. This points to a strong potassium activity. The bottom leaves show a fairly long stalk and a well developed longish stretched blade. Further upwards the stalk becomes shorter and shorter, the blade broadens into an oval shape. Coming nearer to the flower the stalk disappears, the leaf reaches towards the stem.

In the end the stalk becomes lanceolate. It envelops the stem with its finely formed spears. So the leaves show a real metamorphosis. They are quite astonishingly dark green and very smooth, which in a certain sense contradicts the extremely pinnate form. If you taste the yarrow leaves they are very sharp and bitter at the base, and become more aromatic towards the blossom.

Looking at the blossom we see the stem splitting into the umbrella shaped flower. It does not have a very distinctive, out-raying face as some other flowers do. It is rather more concentrated in itself and gives the impression almost of a dried flower. It is white, sometimes pink, in colour, but not very shiny, as if it would hold back some of its powerful force. The whole umbrella is composed of many flower baskets (umbels), which are formed of tiny little blossoms (pedicles). So the yarrow is truly a typical manifestation of the *Compositae* family; its nature reveals an enormous differentiation (sulphur) and at the same time a great concentration (potassium).

One could say we have defined a polarity in the yarrow. On the one hand we have this sturdiness of the stem, with a strong ligni-fying tendency which causes them to last so long (one can often see them standing in the meadows the following year) and at the same time soft, dark green velvety and apparently succulent leaves. On the other hand there is a very highly developed flower and enormous segmentation of the leaves. This polarity of an earthly and cosmic influence can also be observed if we look at the roots. Of course they normally cannot be seen, belonging as they do in the darkness of the earth. But in the last decades much emphasis has been laid on root morphology. The yarrow roots are quite extraordinary, with many thin threads going down very deeply into the earth, like a stream. Again there is an inner activity, a motion, like in the leaves, but at the same time they are so firmly bound in the soil that it is almost impossible to pull up a yarrow plant. So there is the movement of the roots flowing down into the earth, but at the same time a steadfastness like there is in the stem. In autumn and winter, just below the surface, the stolons form runners which then produce new shoots early in the spring. Therefore we often find patches of these stoloniferous yarrows in open pastures.

Potassium and sodium in the yarrow

The yarrow reveals the polarity of earthly and cosmic forces in a very distinct way. Its outer appearance is a remarkable manifesta-tion of the forces which its own being sends into the earthly realm in order to build up, supported by earthly forces and substances, this image of its being. What we perceive is not a reality, it is an image. The forces that build up this image are real, but impercep-tible. They originate in the cosmic and earthly realm and are directed by the true being of the yarrow, which remains in the spir-itual sphere. It is important to realize that what we see is an image, mere semblance. But it points to a spiritual reality which works through two principles. One principle shapes the yarrow according to its cosmic archetypal form, causing that which brings about differentiation of the leaves and flowers, the scent, taste and colour of any of its organs and also its fragrance. We may call this the sulphur principle. On the other hand we have pointed to a diversified root system, the sturdy stem, the strong turgor pressure potential causing succulence of the pinnated leaves very similar to the round shape of the growing point. All these phenomena are caused by what we call the salt or rather the potassium principle. The latter gives the plant its outer earthly appearance, the sulphur principle governs the cosmic archetype they manifest.

Affirming this polarity we may better understand why Rudolf Steiner, when speaking about the herbs that are used for the preparations, speaks about a particular herb in relation to certain earthly elements. In the case of the yarrow he speaks about the potassium and the wonderful, quantitative relationship between potassium and sulphur with regard to protein formation. He refers to this unique content of sulphur in yarrow, in relationship to its ability to work with potassium, as being an ideal model within the plant kingdom.

So we must now ask, what is the nature of potassium, where does it occur? We find potassium as an essential constituent all over the mineral kingdom, especially in mica, feldspar and therefore in great quantities in granite but also widely distributed as a salt. It is quite interesting that the quantity of potassium relates strongly to the quantity of silica in the earth's crust. The more silica there is, normally the more potassium can also be found there. Potassium is really an earthly element. I would call it the representative of the earthly realm altogether. It is, as all physical substances, submitted to physical, calculable laws that we can study in physics and chemistry. If we make an experiment and dissolve a potassium salt, for example potassium chloride, in water, its crystallized state will completely disappear and it dissolves. The property of the water will have changed according to the quantity of dissolved salt. If we then warm the water until it evaporates, the salt is again formed, in exactly the same quantity and crystal structure. The laws of the physical realm are finite and can be calculated. From this point of view it is justified to say the physical earthly realm is dead. It is highly disconnected from its spiritual origin. But potassium stands out through yet another property. It is 0.001% radioactive as it occurs in the earth's crust: it underlies a decay. What does this signify? Potassium is a physical substance at the border of the subnatural world, the subsensory world. From beyond the border there are beings actively exposing the forces of electricity, magnetism and radioactivity to nature. The nature of potassium as the representative of the salt-forming elements can be understood to be a link to subnature, marking a border, as it were.

In contrast to it is sulphur — the substance 'the spirit moistens its fingers with' — the representative of those elements that are the link with the supersensory, the supernatural world, with the cosmos. So we find yarrow is a revelation of a very strong interaction of an extreme polarity. The interacting poles of sulphur and potassium as they occur in yarrow seem to be the fundamental basis for an ideal formation of protein.

Sulphur

Potassium

The yarrow, like any other plant, takes in potassium through the roots. There the potassium as an earthly substance, with all its properties we have described above, disappears. It leaves its physical environment and enters a living one. Of course we may find potassium to a certain extent still in a salty state in the roots. But in the process of growing and developing — first in the still watery leaves and all the more in the successive ones — the potassium becomes estranged from its physical properties and instead takes on properties, step by step, that relate to the living. Within the living sphere it becomes a carrier of living forces. The result of their work in relation to potassium can be seen in the sturdiness of the stem, in the formation of tissue, the scaffolding of the plant culminating in the lignifying tendency, and on the other hand in the swelling, succulent appearance. If you have a deficiency of potassium in soil and plant, on a very hot day the leaves tend to hang down. Evaporation is even increased because the stomata do not close properly. Their regulation and likewise the pressure of the fluids in the cells, the turgor potential, is a function of potassium in the living context. Down in the roots the potassium still has salty properties. The further it comes up through the stem, participating in the formation of the sequence of the leaves, it begins to be under the guidance of the sulphur, and is step by step potentized towards the blossom in the sense of a continuous estrangement from its physical properties, its being bound in space. A real potentization takes place in the process of time, that is in life, as I described in the last chapter. The metamorphosis of the plant is an outer image of this potentization of potassium and its related earthly elements within the course of the full unfolding of the plant. The further up we look in the plant from stage to stage, the more the potassium loses its definable physical properties and becomes a carrier of living cosmic forces, which are unquantifiable.

It no longer marks the border to the subnatural world, it has been lifted towards the border of the supersensory world by the sulphur process. This is the mystery that takes place within the plant, in each different species in a highly specialized way. Yarrow is specially qualified to master the potassium-sulphur process in favour of an ideal formation of protein, the mercurial result of this functional polarity.

Here potassium is no longer detectable in the plant by means of chemical analysis. Of course we can find potassium again if we burn the plant to ashes, but then the plant is dead and what we obtain is salt. We cannot detect the potassium in the very state to which it is raised in the living sphere. The possibility of detecting radioactively marked elements in the living plant is not a counter-proof. It only shows the plant life is submitted to the conditions of its environment, as becomes manifest in the dangerous pesticide residues and other contamination frequently found. Plant life can either be weakened; it is then less able to transform elements from their physically bound properties, especially the subnatural properties. Or it can be strengthened. This is a matter of appropriate manuring: a matter of dealing with the yarrow preparation for instance.

It is impossible to detect the gradual transformation of potassium by the sulphur process through analysis. It comes to the utmost of dilution and potentization, where the plant ceases to grow, where the flower develops. The question is, how are we able to give permanence to this culmination of plant growth, the flowering stage? Because the very moment the flower appears, it fades away. It cannot perpetuate itself, it just lasts a moment; it is a kind of *status nascendi*, the full opening of the flower towards the cosmos.

Looking at the flowering stage of the plant we can make two observations. One is that the being of the yarrow appears as an image in its perfection in the flower, in the shape, colour and fragrance, but in the unfolding of the flower it comes to an end. The plant dies into the flower. There it is touched from outside by its higher being, which was at work all the time throughout the growing process. On the other hand we can observe this gesture of the flower opening towards the cosmos. We must learn to inwardly follow this gesture with our thinking and feeling: we might approach it with an expression of total openness, devotion and willingness. Observing the flower we again meet a polarity. On the one hand the sulphur process, supported by the potassium process coming to an end in space and time in the blossom, leads to a perfect image of the true plant being. On the other hand if we follow the potassium process from the earth upwards, supported by the sulphur process, it begins in the dead state of salt, ending up in this infinite openness and a gesture of willingless will. This gesture of willingless will signifies an embryonic state to which, in my opinion, Rudolf Steiner is referring when he spoke about the plant releasing germinal forces into the etheric world.[2]

Yarrow blossom and stag's bladder

The blossom fades away. Then the question arises, how can we preserve that which has come to an end and that which hides the potential of a new beginning in the blossom? How can we preserve that very moment before it fades away? There are three possible transformations to which the blossom may be submitted. One ends in seed formation. All the processes in the blossom are yet undetermined. It rays out its shining colour, its fragrance and aromatic substances, but the very moment seed formation takes place, all processes are determined to serve the newly formed germ. The seed has just to fall to earth and the new plant will shoot up. The second transformation of the fading flower is when it fall to the ground and forms humus. Rudolf Steiner points to the third transformation, as mentioned above. It is the surplus germinal forces which vanish from the flower into the etheric world and become the substance for a future macrocosm.

Our question is, how to give permanence to the flowering stage before these transformations take place. I think that Rudolf Steiner had this question in mind when he indicates the first step of making the yarrow preparation. The answer cannot be found in the plant kingdom. We must search for an answer in the higher, animal kingdom. We must find a technique, as it were, to lift the flower beyond its natural limitation, over a threshold onto a higher level, onto the level of astral effectiveness, which is present in the animal kingdom. The soul being of the animal has incarnated into space and time. All its organs are formed out of forces which work into the physical and living realm from beyond space and time. The plant is merely touched by such forces from outside; 'the spirit moistened its fingers with the sulphur' and shapes the plant from the outside, while the animal is inwardly permeated by its astral being. Astral forces of the animal sculpt its organs into the stream of life and endow it with the ability to function in permanence. So we must seek on a higher level to find an answer to our question.

We must look for an organization to which the yarrow, as a

medicinal plant, has a healing relationship. Rudolf Steiner said that the yarrow 'can make good all that is due to weaknesses of the astral body.' Its healing force is very much related to all processes that take place in the renal system. In order to preserve the very moment of the unfolding of the yarrow flower, Rudolf Steiner recommended us to take the bladder of a red deer stag. We will see why later. The first step of making the preparation is to gather the flowers and stuff them into the sheath, this spherical stag's bladder. This step is one of inversion for the yarrow. Previously the yarrow flower was open to the farthest cosmos. Now it is exposed to the astral forces that are working into and within this sphere of the animal sheath. Before the yarrow flower exposed itself in a willing gesture, the potassium process was potentized upwards into a germinal state. Now, within the stag's bladder, this substantial germ becomes the carrier of the astral forces which are transmitted and concentrated in the flower by the bladder. A complete inversion is performed. An outside process, which cannot proceed out of itself, becomes an inner process, the beginning of something quite new.

But what about the bladder? It is an organ of concentration, of substances which derive from the inner ensouled life of the animal. It is impregnated with the experiences which the animal had in its soul. They are drawn out of the blood stream by the kidneys and are released and concentrated in the bladder. It is an organ of excretion, very much aligned to the kidneys. The renal/bladder system is related to the most alert of the sense organs, to the eye. This fact can easily be observed, for instance, if you enter a cowshed a bit abruptly. The cows, slightly shocked out of their dreaming perception, stare at you and within a very short time you hear a rush and a 'pat, pat, pat' as they excrete their manure. The cow perceives you, but her response is not an intellectual one. she does not recognize you and respond to you with her head but with her metabolism. Her consciousness is in the rear, where the response from her soul being appears not as a spiritual recognition, because the animal has no ego, but as a physical excretion. Now we see how closely related the kidney/bladder system is to what we perceive with our eyes. The eye is the polar opposite to the kidney in its function. The bladder concentrates substances from the wide range of the inner life of the animal and excretes it to the outside world. The eye concentrates the content of unlimited perception from the outside world and then excretes it, as it were as a picture, into the inner life of the soul.

We take the bladder and the yarrow blossoms, both of which are completely disjointed from their origins. The blossoms are disjointed from the relationship earth-cosmos, from the natural life of the plant, and the bladder is disjointed from the reality of the organism of the animal. The blossoms are internalized, and the bladder, having been inside the animal, now becomes an object in the outer world. They are inverted.

Why do we use the bladder of the stag? Studying the Agriculture Course in this respect we may discover two aspects which we have to distinguish. The bladder, especially of the stag, is an image of the whole cosmos, in its spherical form. So firstly we have to consider the form principle. The other aspect is that Rudolf Steiner refers to the special material substance of the bladder. It is formed out of the inner activity of this very sensory-active animal. A powerful stream of perceptions of the present effective cosmic-earthly environment enters its head and condenses to the material substance of the opposite pole, the metabolic region. Observe any stag. Its nervous alert eyes, its mighty antlers stretching out and growing as long bones through the skull, it is as if the whole being of this animal expands far beyond its head into the bright surrounding world. Secondly, therefore, we have to consider the substance principle. The form is a replica of achievements from the past. All animals are a replica of the wisdom of the past, but in contrast the material substance is formed and built up by forces at work in the present. What happens when we put the yarrow blossoms into the stag's bladder and expose it to the spherical form and substance of the sheath? My personal answer is that the bladder, having been emancipated from its metabolic function in the animal organism, in relation to the aspect of its substance, now becomes a sense organ itself, a kind of eye which perceives, transmits and concentrates into the yarrow flowers, the astral force working in the cosmic surroundings. On the other hand, the spherical form of the sheath endowed with astral forces, enveloping the yarrow flowers preserves them and gives them permanence.

Making the preparations

Proceeding to the second step of making the preparation, we take the spheres and hang them up above the earth, in the air and warmth, so to speak into the belly of the agricultural individuality where the expansion forces are at their strongest. We do this at the time when the elements of air and warmth and the light and warmth ethers are at their most active, in the summer. In the third step we bury the spheres in the soil in the elements of water and earth. We do this during wintertime when the greatest contraction forces are at work in the head of our agricultural individuality and when the chemical and life ether are active, independent of being bound up in plant growth. The present cosmos works vertically above and below the earth in the elements of warmth, air, water and the solid earth, that is in space, and it works in time, as it were horizontally, during the seasons. So by exposing the bladder to these elements vertically in space and horizontally in time, it and its ingredients are endowed with etheric and astral forces of the present activity and revelations of the spiritual world. The flower of the yarrow, having been without will, is now raised beyond its natural limitation and ability to become the carrier of the will forces.

What happens during this second and third step, while the preparations are exposed to the elements? I already mentioned that in the first stage, the potassium is taken up by the root and potentized right up to the blossom. This process cannot proceed any further. The blossom fades away and ends up in seed and humus formation. The first step of making the preparations has been to lift the potassium process, which has entered into a kind of germinal state in the flower reflected in its will-gesture, beyond the threshold onto a higher stage. That happened when it was enveloped in the stag's bladder. It created the potential to proceed to a new beginning. In the second and third steps of making the preparation this mere potential is being fulfilled by the spiritual forces working within the elements throughout the seasons in space and time. This means that the potassium is not only estranged from its physical properties and enlivened by the plant, but it is now open to be endowed with inwardness.

This is my understanding of the aim of all the preparations; that physical substances are enlivened. Rudolf Steiner speaks about the aim of manuring: that it is to enliven the solid earthly itself. This can only be achieved if the physical substance is permanently enlivened. This is not the case in the plant. It dies continuously into form. The plant keeps the substantial process alive by growing on and on and by forming leaf after leaf. This cannot proceed in the blossom. In order to give the life process permanence it must be endowed with inwardness. This is what the animal can teach it. Its soul being keeps its life process streaming.

Evolution into the future

Looking at the three processes of preparation making we can say that the potassium is lifted through three stages of estrangement of its physical properties, from the border of the subnatural, via the life process of the yarrow, to being endowed with forces of inwardness. If we try to follow this thought we might discover what Rudolf Steiner actually means when he discusses the work of the first group of compost preparations, the yarrow, camomile and stinging nettle preparations. He mentions that they are related to one another in generating nitrogen of a yet unknown, completely new quality in the soil. Together they are able to transform the representation of real earthly substances, calcium, potassium and related elements step by step, into something similar to nitrogen, and finally into real nitrogen. What is nitrogen? Rudolf Steiner

points to it being the carrier of the astral forces. In whatever context nitrogen occurs in nature it is related to some astral event, to
forces of inwardness. That nitrogen is an essential element of protein only goes to show that astral forces are already involved in
building up the fundamental living substance. Normally nitrogen
appears above the earth, being the main constituent of the atmosphere (79%). Nitrogen is a dead physical substance whose evolution is complete. It derives from the past, being the carrier of the
Old Moon wisdom. All revelations of nature and its wisdom are
due to the existence of nitrogen. One could say that it is a physical
representation of what comes to us from the past.

But with our yarrow, camomile and stinging nettle preparations
we are producing a new kind of nitrogen, out of a substance marking the border to the subnatural world. We lift them out of their
physical dead state to become permanently enlivened and thus
transformed into a new substance which we may call nitrogen
because it is a carrier of a new kind of inwardness, of an astrality
that works from the future into the present. A new nitrogen is
formed which does not relate to the past but to the future. I would
say that the goal of all the preparation work is to enliven physical
substances to become receptive to forces working from the future
into the present. This opens the gate to a new evolution, or shall we
say involution, into the future.

The only being on earth that is able to have an active relationship to the future is the human being. All other beings represent an
evolutionary end. They can no longer evolve out of themselves. I
don't want to enter the debate at this point about the present materialistic concept of evolution and its reflection in the theory and
technique of gene engineering. The soul body of the animal is more
or less body bound. It lives out its past. The plant derives from the
past, but appears in the present as a true image of the present
earthly cosmic relationship. The mineral kingdom has already
fallen out of evolution in the past. From this point of view it is up
to us, up to our freedom, what kind of future development we will
give to the earth. But when we guide a substance from the mineral

kingdom through the plant kingdom into the astral realm, we enable it to become a carrier of future forces. Calcium, potassium and similar elements acquire an inwardness which, according to my understanding, is what Rudolf Steiner means when he indicates that they are transformed into this new substance of nitrogen. So biodynamic farming actually means that we take over the responsibility for the earth, not by just working in such a way that we continue what is there. This happens in ecological, organic farming. But we need to endow what is there with future forces, so the earth itself can participate in the future development of humankind.

These aspects relate very much to what we eat. If we just eat what is there, the finished process in seed and fruit formation no longer provides adequate human nutrition. Our task is to change the inner quality of protein, this archetypal creation of nature, together with our own development into the future. In order to do so we need to have such preparations, the summit of manuring, by which a new kind of nitrogen participates in protein formation. Therefore quite a new quality of food will come about. This is of great significance, but it is secondary. As a precondition it is more important that we endow nature with evolutionary processes as such.

I am very glad that we have the opportunity to talk about these far-reaching aspects here, seeking for spiritual understanding of what we are aiming at with the preparations, although it is not an easy approach. If at the same time we work with the preparations practically, then we have another sphere of experience. Being involved willingly we create within ourselves the 'soil' wherein these images that we are trying to develop by thinking can germinate. I think it is through this inner relationship that a deeper understanding will increasingly come about. What we urgently need is such an understanding. It is the only source of renewal for the biodynamic work and movement.

6. The Six Preparations

If you read the Agriculture Course it should astound you; the language that Rudolf Steiner uses is quite unique. He speaks to farmers — scientists were not encouraged to participate. Obviously Rudolf Steiner did not really want anyone listening who lived more in the thinking realm than real doers. Ehrenfried Pfeiffer, a chemist, was very keen to go to Koberwitz but was told to stay in Dornach. Many people thought that Rudolf Steiner, when talking to farmers, would speak out of anthroposophy in the style of fairy tales, as it were. It was less than a year since he had given the lectures about the elemental beings,[1] and many people thought that as farming has to do with the elementary world, he would reveal all those beings that are active and at hand while one is working in nature in more detail. But nothing of this. Only once he refers in an oblique way to the elemental beings when speaking of 'the spirits of nature' in the context of the yarrow preparation.

What language does he use? It is a new scientific language which at the same time stimulates will activity. Actually the agriculture course opens up a new chemistry, one of the living. What was its scientific content? I have spoken to quite a few of those who participated in the course in 1924, and they all said 'we almost drank in the whole course ... we didn't understand very much.' I was once told the story of Count Keyserlink, when he was coming down the stairs in Koberwitz Castle with Dr Steiner after the third lecture and Rudolf Steiner kindly asked him whether he had understood the lecture. Count Keyserlink replied 'Not a word, Herr Doktor!' So that was the situation. He spoke in the most modern scientific terms of that time.

Right from the first and second lecture he spoke about silica and limestone, clay and the humus substance in the living earth realm. In the third lecture he spoke about protein and its constituents, oxygen, hydrogen nitrogen, carbon, sulphur. Later he spoke of potassium, calcium, phosphorus, iron, arsenic, lead, and, in relation to the dandelion preparation, an element which does not yet exist in the periodic table. How do we understand these elements which have been separated out of their natural context by modern science. They did not exist as such in ancient alchemy, where the aim was not to define matter but to behold the principle of living processes. In the 1860s, at the height of materialistic thinking, the whole of nature was split into its elements and the periodic table was introduced.

In his lectures Rudolf Steiner did not refer to ancient knowledge but took up the contemporary scientific approach. He spoke in a way that will perhaps be understood in centuries to come. It is spoken in a consciousness which will be the common consciousness in the far future. The participants listening to it were not able to really grasp it consciously, but they had open minds to just receive it. Here in the Agriculture Course we encounter a fundamental law of modern initiation. This does not mean to just renew an old wisdom. Of course anthroposophy gives an understanding of the knowledge of the ancient mysteries, but it does not continue it untransformed. Modern initiation in the anthroposophical sense means to take as the starting point the knowledge of the present time and to extend it to relate to spiritual knowledge. Modern initiation does not take the ideas from heaven and bring them down on earth as in the old mysteries, but it is the other way around. The most advanced materialistic scientific achievements are the starting point for a knowledge of higher worlds.

Science of matter and science of spirit

Rudolf Steiner once described how he was able to write the book *Esoteric Science*, which ought to be familiar to biodynamic farmers. He was thoroughly familiar with the works of the leading scientists of this time: Darwin, Lyell and Haeckel. Haeckel, a brilliant scientist, built up materialism to a world conception of monism.* So, studying Haeckel, Rudolf Steiner recounts that he took his ideas, neither saying they are false, nor right, but impregnating his soul with them. Think of what this means for a man who was fully conscious of the reality of the spiritual world, to freely concentrate on these utterly materialistic ideas.

Then he describes how he conveyed them to the spiritual world. And as an answer from the spiritual world came the contents of *Esoteric Science*, the spiritual evolution of earth and humankind. In a similar sense I am convinced that the response from the spiritual world to the purely materialistic orientation of the nineteenth and twentieth centuries is the invention of the biodynamic preparations. It is as it were the answer to the dissection of nature into periodic tables.

Rudolf Steiner was well informed about what was going on at the top levels of science. In the course of the twentieth century, materialistic science created modern agriculture by reducing all natural processes to the level of the elements as a common denominator. Soil fertility became a matter of a surplus or deficiency of elements. The single element, for instance nitrogen, became a measure of soil fertility. If we see it is lacking through analysis, we must add it. Hydroponics is a subsequent development of this way of thinking. The exact composition of elements is worked out by computer in relation to the needs of the developing plant, and according to the inorganic concept of cause and effect, the plant delivers a pre-calculate yield. The materialistic concept of nature has been transformed into a mechanistic technology. Nowadays manuring usually means to take single elements and feed the soil.

Rudolf Steiner earnestly takes his starting point in modern science, but develops it in another direction in the Agriculture

* Ernst Heinrich Haeckel (1834–1919), German scientist. His most popular work, *The Riddle of the Universe,* explains the universe as brought about by purely natural causes, without the intervention of any divine power.

Course. There he refers to the compost heap and speaks about a different level of reality, pointing to the origin of organic matter coming from the plant kingdom. This organic matter decays and living forces are released. That is one level of manuring. Another level is the ensouled matter that derives from the animal kingdom. Cow manure for instance is not simply a composition of a wide variety of elements. Its unique manuring power derives from the ensouled forces it releases. If we work on the mere inorganic level with artificial fertilizer, we are able to calculate their effect. The matter and its properties, the electromagnetic forces, are in balance. In the living sphere the balance is shifted towards the living forces. In relation to artificial fertilizer the composted manure is more effective that the mere matter it is composed of. There is a surplus of living forces in relation to the amount of substance you add to the soil. On the level of ensouled matter, for instance, cow manure is the highest state of manuring power in nature. There the manuring effect is far higher than the amount of material components would suggest. Now one of the fundamental questions of the Agriculture Course is whether there is a manure that contributes even higher forces in comparison to the amount of ensouled animal matter? What may our contribution be from the spiritual level as human beings? The answer is the preparations.

The six preparations

I would like to refer to the preparations from a special point of view, summarizing how the six preparations are related to one another. Is there a deeper interrelationship in their sequence?

Considering the relation Rudolf Steiner points to between the herbs and the physical elements they are specialized to work with, we have seen that yarrow is highly capable of dealing with potassium. Potassium forms salts which are pure matter. Within the mineral realm it somewhat marks the border to the subnatural world because of its radioactive properties (0.001% of the potassium in rocks is radioactive). So the yarrow really has the capacity to adapt

An oak tree

and work with an outstanding representative of the earthly realm altogether. This happens with the help of sulphur.

Camomile works with and is able to overcome the physical state of potassium and calcium. Calcium is a substance which is not a main constituent of the depths of the earth. The limestone deposits are to be found nearer to the earth's crust. The calcium activity has very much to do with sucking in the forces of the inferior planets (Venus, Mercury and Moon) that are working above the earth. Calcium therefore has a closer relationship to what is immediately below the surface of the earth.

The stinging nettle works with potassium, calcium and iron. In its working it comes nearer to the earth's surface. It is the stinging nettle, together with the camomile and yarrow preparations, which introduce what Rudolf Steiner calls *nous* to the earth. That is, organizing the soil processes as if they would be governed by a higher organism. So these three preparations are active in vitalizing the earthly realm itself. The oak bark is characterized by the quantity of calcium distributed in it. It has passed through the life processes of the oak and has been excreted into the bark. This signifies its special quality. It is a different quality of calcium to that below the earth.

The oak bark is formed above the earth, the oak being turned up earth, so to speak. Due to this special quantity and distribution we use the harmonizing and healing forces of the oak bark where the earthly element of calcium in its working is just as near to the soil from above as the stinging nettle is from below.

Next in the sequence is the dandelion which has the capacity to produce a substance which is not mentioned in the periodic table. It relates the silica which is distributed throughout the atmosphere in homeopathic doses, to the earthly potassium. Rudolf Steiner calls the dandelion 'a messenger of heaven'; that is, a mediator of the forces from above and not from below. Dandelion has the capacity to relate a cosmic substance, the homeopathic silica, to the earthly element, potassium. The main activity is to draw forces in from above.

This is even more the case with the last in the sequence, valerian. Its function is to be a mediator of mainly warmth suspended in the air and the etheric forces beyond. It is active at the borderline between nature and supernature.

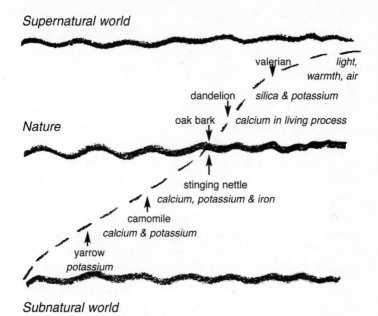

Supernatural world

valerian — *light, warmth, air*

dandelion / *silica & potassium*

oak bark / *calcium in living process*

Nature

stinging nettle
calcium, potassium & iron

camomile
calcium & potassium

yarrow
potassium

Subnatural world

Looking at the relation of the different elements to the different plants we can see the more we come towards the valerian, the working of the supernatural to the earth dominates, and the more we come towards yarrow the working of the earth's depths is mediated. The polarity of the forces of light and darkness is balanced out. From the point of view of the agricultural individuality the first three relate to its head, promoting its sensory activity towards the middle, while the other three are working from its 'belly' to activate the metabolic forces of will power towards the middle sphere.

Sheaths of the preparations

This polarity in the sequence of the six preparations can become even more visible when we look at the sheaths.

The yarrow flowers are encapsulated in a stag's bladder. The bladder manifests the end of the metabolic processes, concentrating what is excreted to the outside world. Camomile flowers are filled into bovine intestines, which mark the beginning of the metabolism, where digestion takes place. The stinging nettle is related to the rhythmic system, to the heart. The stinging nettle does not require an animal sheath, because it has such strong astral forces. Normally the astral being of a plant is itself reflected in the flower. Looking at the shape and colour of a lily or rose, for example, we are deeply impressed by this revelation of astral forces. The stinging nettle has a very insignificant flower which is quite hidden in the foliage of the upper part, but the whole plant can be perceived as being a flower. The stinging nettle does not impress the eye so much as the sense of touch. We feel this burning intensity. It is a different kind of revelation of the astral to looking at the colour and shape of the flower. The flower leaves us free in our feeling observation whereas the nettle directly affects our will and feeling body. It relates to the heart. Rudolf Steiner indicated that the stinging nettle should envelop the heart. So its inner nature is to envelop and organ, rather than the other way round. Of course the heart has small inner spaces but one cannot really fill these cavities; it is not a real sheath. The heart has an inner and, in connection with the circulatory system, an outer activity. So too the nettle. It is just as active to the outside, for instance guarding itself with its stinging silica hairs, as to the inside, harmonizing itself by its strong healing capacity. Its astral body has permeated the whole plant and provides its own sheath.

Considering the oak bark, and the skull into which it is filled, we proceed to the sensory pole of the animal, the head. We seem to have reached the end. How can we get any further than that? The dandelion is enveloped in the mesentery, which seems to be a turning back to a metabolic organ. But is it really a reversal, or a continuation towards a more highly developed 'head'? If we develop our astral body it is transformed into the spirit-self. The cow, from which we take the mesentery, has no spirit-self, it has no ego. But it is endowed with a very highly developed astral body which is open to its group soul. The head is not endowed with it. Towards its head the cow becomes dull. On the contrary, this uniquely

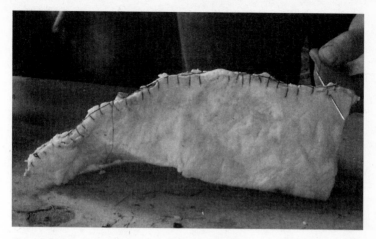

Dandelion preparation sewn into a mesentery

developed astral body of the cow is concentrated in the mesentery which envelops all the inner organs of metabolism. The mesentery is interwoven with nerves. The solar plexus is a part of it. This signifies that the cow perceives something through its inner nature, not primarily through its head, about the inner nature of matter which it analyses while digesting it. When the cow takes in food and chews, a most intensive living activity of perception takes place, especially by the life sense, via the mesentery. In this context the significance of the cow's horns can be fully understood. The horns, situated on top of the head, reflect the immense etheric forces which are released by digestion and which steam up into the horn bone. These forces are reflected back into the 'inner heaven' of the mesentery. Just as we look up into a starry heaven, so the cow looks into her own inner heaven where she perceives by reflection what she has taken in as food from the outside world. Through the digestion of fodder she perceives the outside world and thus the inner nature of matter is revealed. It is for this reason that we can look at the cow's mesentery as a more highly developed sensory organ. the mesentery is really an appropriate sheath to preserve and maybe even enhance this wonderful capacity of the dandelion to be a messenger from heaven. Thus the cow mesentery is not a metabolic organ but a sensory organ which continues on from the skull preserving the oak bark.

Proceeding to valerian we see it is not like the stinging nettle which has the unique capacity to envelop itself. Valerian is enveloped by the cosmic periphery. It forms a sheath of warmth wherever we spray it around the compost heap, or maybe around the crown of a tree. It may be regarded as a continuation beyond the working of the dandelion.

So looking at the plants and their sheaths we have found an inner relationship in their order. If we consider the working of the preparations, the same order in their sequence can be observed. The yarrow, camomile and stinging nettle preparations form a unit. They work from beneath the earth, transforming dead solid earthly matter and so enlivening the soil. Oak bark, dandelion and valerian also form a unit. They enhance and enliven the plants themselves, drawing in forces from above via the manure. The oak bark makes the plant resistant to diseases. The dandelion enables the plant to become sentient for the earthly substances it needs. Valerian surrounds the plant life with a warmth mantle.

The human contribution

To conclude I would like to ask what is the spiritual contribution of man in working with the preparations? Because nature provides what we have spoken about so far as nature's deeds. Nature provides all the components of the preparations. The first step of our contribution is to grasp the idea underlying the preparations. They are the result of spiritual research by Rudolf Steiner and can only be comprehended in the context of the whole of anthroposophy. Comprehension in this sense means to deepen our understanding of the spiritual insight that the yarrow has a particular function. It is placed in the bladder and finally exposed to air and warmth during summertime, and water and solid earth while buried in the soil during winter. We must open our minds to the background of these ideas. We must come to an understanding of the beginning, the

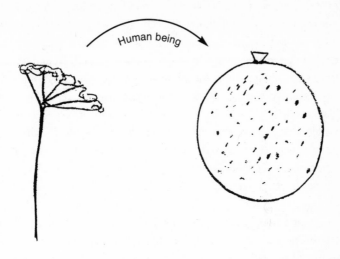

alpha, of the evolutionary process of the yarrow, or the bladder. We hold the end of the evolutionary process, the omega, in our hands. The underlying idea to relate the components of the preparation to each other derives from spiritual research which reaches into the realm of eternity where the beginning, the alpha, and the end, omega, are joined. This requires a continuous effort to study the Agriculture Course, having as a background the whole of anthroposophy, not merely reading it but weighing it word by word and entering the sphere out of which these words were spoken. We must learn to penetrate beyond the spoken word into the realm of imagination and inspiration. To widen our spiritual cognition is one challenge. Through this spiritual understanding our will is activated. But by mere spiritual recognition no yarrow preparation will ever come about. We have to actually make it! But what is our contribution in this process?

It is to lift the yarrow flower, which has come to an end at this stage, beyond its evolutionary reality. This happens when we put it into the bladder, into a higher reality of inwardness provided by the astral of the stag. We are active in building a bridge from here to there. We stand with our will guided by spiritual insight. Nature provides the past, she is not able to build this bridge. Our will and our insight provide the future and therefore we are able to combine what has developed separately. The next step is to hang the bladder above the earth and expose it to warmth and air in summer. The third step is to bury it in the soil in wintertime. Again it is our activity and insight to link what would never come together by itself. This work enables us to become increasingly involved with our whole being in this process. We deepen our feeling of what happens during summer and winter on our farm at the very place where the preparations are exposed to the seasonal influences. Through this activity of thinking and will, we might be stimulated to sense with our feeling those spiritual beings active in winter or in summertime. We are conscious, for instance of the indications of Rudolf Steiner about the work of the archangels.[2] Gabriel is working from beneath the earth in summertime developing nourishing qualities in fruit formation while Uriel is governing the realm above the earth, judging what is right or wrong. This spiritual activity of the archangel sphere is the inner content of what we experience as light, warmth and air during summertime.

These filled bladders become sense organs for the astrality that lives within light, warmth and air during summertime, and when they are buried in the earth in winter they then become a sense organ for those forces at work there at that time. In winter the archangels work in the opposite way: Gabriel works from above in his loving gesture and Uriel, working from beneath the earth, stimulates the force of thinking. When we expose the preparations to the outside light, warmth or air we expose them in reality to forces working therein, and we prepare the preparations in such a way that they are receptive. It is through our activity and free will that the preparations become receptive and are carriers of these powerful workings.

I wonder if you have ever thought about what actually pushes you to work with the preparations? You might have become interested by reading some guidelines and decided to start working with them more or less through confidence and belief. But step by step, in making and applying them the inner urge arises to work towards an understanding of their spiritual origin. The more you develop this spiritual activity while doing it, the freer you become and the less stimulated you are from the outside. The more we understand what we are doing the closer we come to self determined freedom.

When we dig up the preparations there appears a humus-like substance, yet almost nothing substantially. We take a tiny portion from this and put it in holes we have made in the compost heap. We cover it over and leave it. Powerful, harmonizing and transforming forces unfold and govern the decaying matter. Beyond observation something happens while quite hidden within the heap, the result of which is spread out on the land. Have you ever thought about the fact that Rudolf Steiner talks about six preparations. Why not seven, if you think about the significance of the seven planets for instance? I have come to the understanding that there is a seventh. It is the developing soil itself, now enlivened in its fertility by these six preparations via the composted farm manure being spread over the whole farm. It is our aim, and Rudolf Steiner describes it in the Agriculture Course, to enliven the solid earthly element itself. The soil can be viewed as the seventh where all six preparations are combined in their working and develop the seventh, the diaphragm, which is the idle sphere between the head and belly of the agricultural individuality. Our contribution in this unique process is to transmit one stage to the next. This activity requires our thinking just as much as our will. In between, our feeling is the gate through which we learn to understand what comes to us by spiritual insight and by exercising our will. This opens a space for us to act in freedom in nature's realm of necessity. The highest we may contribute to our natural environment in the future is this freedom born out of spiritual insight.

7. The Biodynamic Preparations as Sense Organs

First of all the question I would like to deal with is, What is the reality of these preparations? Looking at the world, the only reality we are really fully aware of is that of ourselves as ego beings. The human being is actually a citizen of two realities which are concealed from him. There is on the one hand the spiritual world into which he reaches with his own spiritual being, and the earthly physical world in which he stands with his physical being.

Spiritual world	will
human beings	feeling
earthly realm	thinking

Human thinking, feeling and will

We relate three soul faculties, thinking, feeling and will to these two poles. These three abilities are still very much underway in their development. In our will we are in contact, although unconsciously, with the spiritual reality by submerging into it our everyday work. Also we enter the spiritual reality when we fall asleep every night. This unconscious encounter during our sleep enriches us with new impulses. The counter pole is our thinking, where we are really awake and conscious but disconnected from the spiritual. We only touch these realms by thinking. What enters our consciousness is the sun illuminated surface: the outer image of the creative being behind. We develop an onlooker consciousness. We are not able to penetrate the outer appearance. In the middle our feeling is active, half consciously, in a more dreaming state, merging the two poles and awakening thereby our self consciousness. That is the situation of the human being nowadays.

The present constitution of the human soul is such that these three faculties are merged into the unity of our conscious being. But we are passing a threshold to a further development of these abilities. In beholding ourselves we can become aware that we think something most clearly and yet may do the exact opposite! For instance we drive our cars knowing the damage they do to our environment. Or there are large numbers of people deeply involved in scientific work experimenting with genetic engineering, who are completely disconnected with the reality of where their results are put into practice. So we seem to know almost everything except this one small part: what we are actually doing. And the feeling life is left alone in between, separately nurtured by the media, for example. Passing the threshold I mentioned earlier, these three soul qualities become more and more independent. The more they fall apart, the more we are challenged to strengthen our ego-being to consciously harmonize these diverging soul abilities. In so doing we step onto the path of inner schooling towards higher knowledge. That means we learn to feel the dark and light aspects of our thinking, and develop the faculty of imagination with our thinking, inspiration with our feeling and intuition with our enlightened will. This marks the direction of recognition into the future.

These three faculties can only be developed when these three soul abilities become independent in the soul and are governed by our ego being. Rudolf Steiner developed these faculties to an advanced extent and thus became a researcher in the spiritual world. He can be regarded as being the modern representative of humanity who developed, independently of the physical body, receptive organs by which he was able to sense and become conscious of the different regions of the spiritual world. By spiritual research the preparations have come from these realms.

When Rudolf Steiner came back to Dornach from Koberwitz he gave a lecture reviewing the Agriculture Course and mentioned that 'Anthroposophy relates to the highest esoteric and practical.'[1] And truly in the Agriculture Course the highest spiritual reality is combined with the most practical life in the physical world, the result of which is the invention of the preparations. We can look upon the preparations as being very closely related to the human being, as the three most advanced faculties of recognition radiating into the spiritual world have brought them about. As Rudolf Steiner widened his being in full consciousness to the polarity of both concealed realities he was able to create a new technique in the living sphere which is justified by the truth endowed in it. With our normal onlooker consciousness we deal with the preparations and recognize that they do not correspond right away to our thinking, feeling and will.

Our inner work gathering the preparations

So working with these preparations we must become aware that we must work on our thinking, our feeling and our will at the same time. The first question we have to pose is, are the preparations simply recipes out of anthroposophy? Don't they work for themselves once they are made? Is the human element an essential condition for their working? In attempting to answer we must first say that the spiritual concept of making the preparations cannot be found in nature. You might observe nature as precisely as possible but you would never discover the ideas to make preparations; they are not inherent in the wisdom of nature. They derive from spiritual insights resulting from research into the spiritual origin of nature which has developed further since it has physically come into being. Something new is added by the human element. I would like to put our relationship to the preparations into an image. It is like a painter who intends to paint a picture. He takes his canvas, his palette and his brush and starts painting. After a while, when it is finished a visitor comes along and looks at this picture and, although he hasn't created it himself, he may be deeply moved inwardly simply by looking at it. When we make the preparations we have to consider two steps. We make them; that is like painting the picture. We create a reality. Then we take this picture and offer it to the beings active in nature and they, although hidden, look at it, so to speak, and are moved by the moral perspective giving a new direction to their activity.

First of all we can study word for word how Rudolf Steiner introduces the preparations. We have to do it as regularly and intensely as possible. In the end we ought to know the Agriculture Course by heart. So our first approach to the contents of the Agriculture Course is through thinking; we ought to think how the words and descriptions are linked to one another, having been spoken out of imagination, inspiration and intuition. If we take for example the dandelion preparation, we ought to find out, through thinking about it, why Rudolf Steiner speaks about the dandelion as being a messenger from heaven. Try to picture the relationship between the silica in the cosmic environment and the potassium process streaming upwards from the earth via the root. Think over the relationship between the flower and the mesentery, and how the finished preparation makes the plant sentient. Take the words the spiritual researcher uses as a phenomenon. The phenomena of Anthroposophy are the words which the thoughts are clothed in. We meet these phenomena with our thinking in the same sense as we meet the sensory world by natural science. The spoken word in the Agriculture Course is an outstanding expression through which the spiritual world appears to us as phenomena.

In a second step we must try to learn to feel beyond the words into the realm out of which they were spoken. We must try to feel how the indications are related to one another and thereby feel into the sphere of intuition. So we might build up ever and again an inner picture which becomes more and more alive and engenders the enthusiasm to really get down to work and make the dandelion preparation, for instance.

In making the preparations we are not primarily involved with our thinking, but we engage our will by going out at the end of April to collect the dandelion flowers in a basket. Wandering over the pasture, we might just stand still for a moment and dwell on the situation we are meeting there. Try to feel the mood, the springtime atmosphere, this bright yellow-spotted pasture with the urgent green growth underneath. Try to feel how this single dandelion plant, ever since last spring and throughout the summer, autumn and winter, collected from heaven and earth enormous powers. It has concentrated them to lift up this stalk for a moment and open its flower so we can gather it.

Try to observe all the details you have observed while gathering the flowers. Take a spade and have a look at the roots, how they grow down to seek out the potassium in the salty earthly realm, and try to picture how this process is transformed in root and stalk in the white sap pressing upwards. Try to picture how the leaves are shaped and arranged in a rosette. Observe how, during the course of the year, perhaps over a hundred leaves are formed. See how such a plant develops within its rosette a metamorphosis towards the flowering stage of the shaped leaves. Be aware how overnight the plant lifts up a second storey, a loft of single florets that form this one bunch of the dandelion flower. And finally admire how, in the space of a few hours, the flower bases bend down and form the sphere where the seed appears ready to be carried aloft on its little umbrella.

So while standing still there we can build up an image through thinking, through feeling and by being willingly involved. The outer apparently simple deed of gathering the flowers is accompanied by building up this inner image which may become ever richer.

Having dried the flowers in the loft, the first stage of preparation making follows. We take the dandelion flowers and wrap them in a mesentery. We sew it up into a parcel. In this step we are fully challenged humanly, because it really becomes a free deed. Nothing pushes you to do it. It is simply a spiritual insight which we have internalized and work out of. It seems so easily done and not especially significant. And yet what are we doing? The dandelion flower is led beyond its natural destiny, which is seed formation. It is combined with something beyond, a sheath from the animal kingdom. Spiritual insight has led us to use a mesentery. We normally obtain one from an abbatoir, but ideally we would slaughter a cow ourselves. We do not do so, because we are not qualified to, but we should ideally be present at this event and observe the incredible beauty of the inner world of organs when the dead cow is opened up. It is as if the curtains of the astral heavens are opened. Experience the colours there and smell the inner astrality that is suddenly released of such an animal. We can try to sculpt these wonderful forms in our minds and behold their profound composition. We can take out the mesentery and see its transparency. Thus we get a true picture of the mesentery in the context where it has been formed in the animal.

Accompanying the process

Then we take the second step of preparation making. We dig a hole and bury these parcels of wrapped dandelion flowers, covering them with a layer of earth. Again it seems an easy deed, and yet it is a tremendous challenge to our will because nothing forces us to do it. The more we enter this step with our understanding the more our free will matches this challenge. In the second step we not only combine elements of the plant and animal kingdom, we also relate them to the mineral realm. In doing so we lead our thinking and feeling into our will. We search for an appropriate spot, maybe at the edge of the pasture, in the orchard or garden, and there in a sheltered spot we dig the hole. We put the parcels in the hole. Before covering it maybe we put some elder branches over the dandelion parcels so that the animals won't get to their wonderful contents. After we level it up with earth we stand still and dwell upon how the atmosphere is on the very day we do it, maybe Michaelmas, or some sunny golden day in October. Try to feel the autumnal atmosphere. We try to sense that this deed can be a fulfilment of celebrating Michaelmas. We might feel that our deed stems from a spiritual concept which has been brought down by spiritual research out of the realm of Michael into the earthly realm enabling us to make these preparations. Michael governs this spiritual concept. He shows us with his directing gaze the path to follow to put it into reality and leaves its realization to our freedom. He delivers the concept.

At this stage of our work we try to feel such an image and at the same time we observe the conditions in the outer world, of the earthy, sandy or loamy nature of the soil. What kind of rock underlies it? Is it sandstone for instance, or limestone or granite? We must feel right down in the darkness of the solid rock that enables us to stand up. Furthermore we should picture the situation of the surroundings; perhaps there is a fruit tree nearby, a hedge or a garden wall. In winter we might walk across the farm land and come to this spot. We remember what we experienced in autumn and compare it to the frosty atmosphere we are now in. Perhaps snow covers the earth, and crystallizing forces penetrate the earth. We try to imagine how are the forces relating to the dandelion flowers wrapped up in the mesentery. Thus we accompany the preparation process throughout the year and, by building up these images we enlighten our will and our thinking pictures the surrounding while our will is active. Thus the image is more enriched and can be sensed in its spiritual reality.

In springtime, when we dig up the dandelion, we try to picture the situation that we encounter once more. We perceive the colour, smell and structure of the dandelion remains and compare it to what it was like in autumn. Again we stand still and dwell upon the unique phenomenon which we have created. What we have is a completely newly formed substance. It is humus-like and thus belongs to the earthly realm and yet it has become a carrier of a spiritual reality. What we have in hand is a germ in substance and force, wherein the cosmic and earthly realities are united but yet concealed like in a seed. We are the creator of this new kind of seed. It is completely different from, and even opposite in its function to, a plant seed.

A plant seed contains, impregnated in its mineral substance, the cosmic form of a plant. This form is present as a spiritual reality in the seed as a germ, as a potential. When the plant seed unfolds, this potential is determined into this specific physical form. It becomes a defined outer image of its cosmic reality. The seed we are producing by making the preparations is different in its germ function. It does not contain a cosmic reality as a potential, an ideal form which may then unfold into an outer image. It contains a potential to germinate as a substance. It is the potential to enliven the earthly itself. When the preparations are placed in a compost heap or manure heap, they unfold a germinating force in the physical realm. What unfolds is not an outer image. It is the cosmic reality we are dealing with which opens the mineral world up to the processes in time.

A deed for the future

The second step I was speaking of is applying the preparations. We are deeply involved with our thinking, feeling and will, and must learn to harvest all the fruits of our efforts and of our experiences over time. We take a tiny little amount of this newly formed substance, put it into the compost or dung heap. We try to build up a picture of the decaying matter, of the microbial world being active in destruction. In this decaying process which tends to end up in mineralization, the preparation now radiates forces that stand against this decay and have the capacity to transform the manure heap to a kind of organism. The decaying forces are thus balanced out by the working of the preparations. When it is ready we take the compost or manure and spread it onto the land, the diaphragm of the farm individuality, the soil. So we go out in the field and observe how the manure has now been worked in and try to imagine how, from above and beneath, the spiritual reality is uniting with the middle sphere. We must open ourselves in our inner being to the reality we are creating. Don't just leave it and let it happen in the sense of cause and effect.

If we accompany this process consciously all the time we really do produce a work of art. We are painting a picture which is perceived by the beings active in nature. Until now we have only followed the steps of how the dandelion preparation comes into being. But try to imagine the richness of all six preparations and the manifold interweaving images we may perceive throughout the year while making and applying them. When we penetrate all our work with these images we may become true biodynamic farmers. We are obliged to do most of our daily work with machinery, but working with the preparations we have a space within our activities where we really can expose our free will and are able to converse between our inward being and the outside world. With this attitude we overcome the onlooker consciousness that sees it as a matter of cause and effect. But working with the preparations, building up this inner richness, we eventually behold ourselves as a creator of what happens to the interior of nature. We recognize it as our deed relating to our judgment. With this 'seed' we endow nature not only with a substance, but also with the devotional consciousness of our spiritual being.

I want to refer to one remark Rudolf Steiner made in a lecture given in 1911.[2] There he said we will see in the course of the twentieth century and all the more in future the decaying of the earth, of all our natural environment. He continues by saying that we will be very sorrowful about it, and yet when we are really aware of this decaying world we will see that since the beginning of the twentieth century all of a sudden here and there something refreshing will spark from it. And finally he mentions those who are able to look into the background of this appearance can recognize new elemental beings which will become the servants of Christ. They come into existence through a new relationship we seek to build to the decaying earth.

With this in mind imagine what we are doing in making and using the preparations. In making them we combine out of spiritual insight what is separate in nature. When using them we implant in nature a new material, the composition of a new quality, which cannot be defined. The wisdom which reigns in nature is finite and somewhat defined by the ruling laws. But the quality of the matter we are now implanting is the most irrational. It is in its essence related to the quality of love: love is nothing but enlightened self-less will, the inauguratory substance of all future development. Love has the virtue to redeem, heal and transform all social distress. Why should we not be able to externalize this virtue and produce a force of irrational quality which redeems, heals and transforms what is in a stage of natural decay in nature? While working with the preparations I think we have to bear in mind this picture that we really do plant something into the earth that she longs for, but cannot create herself. It needs the free deed of human beings.

Six preparations

Now I want to look at another aspect. We are dealing with six preparations.

valerian
dandelion
oak bark

_____

stinging nettle
chamomile
yarrow

If we follow up the sequence as I pointed out in the previous chapter, we find that the first three and second three preparations belong together. Seen from the earth's surface the first group works from below upwards, the second from above downwards. They form a polarity and in the midst is the diaphragm, the soil, where the plant sends roots downwards and the stem upwards. The third preparation of each group, the stinging nettle and valerian, don't require a sheath. All the others require animal organs as sheaths, each of a different kind. If we look at the sheaths from the point of view of their sense activity we also find a polarity.

Bladder and intestines

In describing the first, the yarrow preparation, Rudolf Steiner mentions the bladder of the deer — normally the bladder of a stag is used — to be the appropriate sheath to envelop the yarrow flowers. The bladder is a concentration organ and at the same time an excretion organ. Rudolf Steiner refers to the bladder saying we have to consider the form aspect; 'The bladder is almost a true image of the cosmos.'

The second aspect is the substance of the bladder. It is built up in three layers. The inner layer is the mucous membrane which secretes slime to recreate the wall which is continuously attacked by the urine, especially the uric acid. It is therefore a layer with an active metabolic activity. A second layer consists of very thin muscles which rhythmically contract and expand. This layer represents the rhythmic aspect. The third, the outer membrane belongs to the serosa which is part of the mesentery, this skin that lines the abdominal cavity and envelops all the metabolic organs. It is a sense-membrane and represents the sensory pole within the three-fold structure of the bladder skin. So when we put the yarrow flowers into the bladder they come in contact with the inner, more lively metabolic membrane. The outer layer is exposed to the elements and the working ethers when the balls are hung up in the air in the summertime and buried in the soil in winter in the watery, salty earth below. Its sensory activity relates it to the outside world and whatever works there is now transmitted to the inner, to the yarrow blossom. It is concentrated there because it is preserved by this metabolically very active inner layer. The blossoms come into the inner, concentrating and preserving realm while the outer, sense active layer is exposed to the forces and substances of the outside world.

Portion of longitudinal section of small intestine

The sheaths of the next, the camomile preparation, are the intestines. Their skin shows the same threefold structure as the bladder although being an organ of digestion it is completely different.

We also must distinguish an outer layer which is part of the mesentery, the peritoneum, a mediating muscle layer, which causes the peristalsis: contraction and expansion. Finally there is an inner layer which consists of innumerable villi giving an enormous enlargement of its surface. Look at a villus, it has a wave-like skin and therefore again provides an increase in the active surface. The villi have a double skin of cells through which the completely broken-down substances of the intestinal tract are absorbed and carried away. Embedded in the mucous skin between the villi we find many lymph nodes and glands.

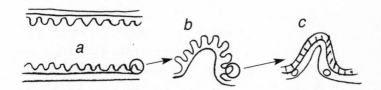

(a) portion of longitudinal section of small intestine
(b) enlarged villus
(c) enlarged portion of villus

In this inner layer of the intestines the ether body unfolds an immense activity by excreting fluid into their cavity. This fluid has the capacity to break down the food completely, right down to its mere mineral components. Although the threefold structure of the intestinal membrane resembles the make-up of the bladder it is opposite in its function. The bladder is an organ of excretion at the end of the organism. It concentrates and releases what has become useless to the life processes. In contrast, the intestines are the entrance where the food is taken in and actively broken down. It is dissolved to enhance the life process. The intestinal tube, through which the food passed and was digested, we fill with the camomile flowers. They come into close contact with this very active mucousy intestinal wall. Again we have the phenomenon that the flowers are completely incorporated in this metabolic realm, while the outer 'sensory' layer is in direct contact with the surrounding solid earth when the intestinal sausages are buried in wintertime. The rhythmical middle layer mediates what is sensed from outside to the interior, where it permeates the camomile flowers and is thus preserved.

The nettle

Considering the sheaths of the first two preparations we move from the bladder, an exit of the metabolic system, to the intestines, an entrance. When we come to the stinging nettle preparation what sheath would be appropriate? We must step beyond the threshold of the diaphragm into the rhythmic system, the realm of the pulsating heart and the breathing lung. But heart and lung are not organs that can be used as a sheath. Now Rudolf Steiner makes two remarkable comments about the stinging nettle.

On the one hand he points out that in making the stinging nettle preparation no sheath is required, and then describes how the stinging nettle 'should really grow around man's heart, for in the world outside — in its marvellous inner working and inner organization — it is wonderfully similar to what the heart is in the human organism.' We must not only be aware of the words with which Rudolf Steiner pictures the results of this spiritual research, but learn to listen and feel behind these words, in order to deeply grasp the idea in its spiritual context. If we try to imagine and feel this picture of the stinging nettle growing around one's heart we may discover a most remarkable truth. We may grasp the idea that what actually envelops the heart of the human being is the peripheral circulatory system. The arterial blood stream into the periphery is then transformed to venous blood and streams back. It is most remarkable that the circulatory system evolved before the heart did. The heart is a later evolutionary development. If we take this picture seriously we could say that the stinging nettle should be enclosed by the whole circulatory blood system. But this is of course not feasible.

On the other hand Rudolf Steiner mentioned that the stinging nettle does not need a sheath. It has the capacity to cover itself as it were. If we study the stinging nettle we will find that all the leaves are covered with tiny silica stings which break and release the stinging histamine and formic acid when you touch it. Towards its periphery the stinging nettle is strongly astralized. It has the capacity, through this outer silica coat, to perceive what is working in the earth during summer and wintertime while it is buried there. It provides its own sheath to perceive and to preserve what is concentrated in its 'inner organization.'

Skull and mesentery

Now we have considered the first group of three preparations, which work from below the earth to enliven, to refresh the soil and to sensibly organize its processes. With the second group we pass the threshold of the soil. They work from above. The first of these is the oak bark preparation. We take the oak bark and put it in the skull of a domesticated animal. So regarding the sheath we now pass from the rhythmic to the head or sensory system. Why do we take the skull from a domesticated animal? The latter can be distinguished from the wild animal because in its physical and mental being it is kept back in a more embryonic state. A domesticated animal does not develop right into the wild. When we approach it, it does not run away, it comes towards you. It is ever and again a most remarkable experience, which seems so obvious, to enter a cowshed and see all the animals standing or lying there expecting you to come to milk, feed and care for them. Their being is open towards you, to your guiding ego and towards their group soul. Looking from a higher aspect domestication means that in former times people were able to keep the animal back in its evolutionary development — manifest in morphological and physiological features — and thus to open its soul being to the group soul and to the guiding consciousness of man. This is the reason why I think that the skull of a domesticated animal is used for the oak bark preparation. The crumbled oak bark is filled into the cavity within the skull.

The interior of the skull is lined with the bone skin. In contrast to the belly sheath we have been speaking of in which the inside was lined with a metabolic layer we find the bone skin in the skull to be a sensory membrane. It encloses the brain and reflects and concentrates into it all the forces of inner and outer perception that constitute the specific animal consciousness. Without this sensory sheath they would tend to get lost in the void. But by being reflected into the inside, the animal is able to unfold its conscious soul being. Now in place of the brain we put the oak bark inside the skull and submerge it into decaying organic matter in a vessel or alongside a brook and let rain water drain through it throughout wintertime. The abundance of disordered etheric forces, which are released by this watery decaying organic matter, have an affinity to lime. They are absorbed by this well-structured calcium skull bone of the domesticated animal, sensed by the bone skin and then reflected and concentrated in this uniquely plant-born calcium structure of the oak bark. What happens is exactly contrary to what we have observed in the metabolic realm. There the outer layer of the organs had a sensory capacity. The inside layer is metabolically active preserving organic residue (bladder) or enhancing organic decomposition (intestines). In the middle is the rhythmic layer of the expanding and contracting muscles. Putting the skull into water, we arrange an environment of an active metabolism outside. Moon forces absorbed by atmospheric water permeate organic matter. Inside we have the sensory active skin which transmits to the oak bark what the middle layer, the skull bone mediates from the outside.

Now we step forward to the dandelion and to the question, What kind of sheath is left in the sequence we have discussed so far? Can there be a further intensification of a sensing capacity in another organ beyond the skull? There is none. We have reached the forepart of the animal. And yet when we consider the ruminants there is an intensification. We find it by returning to the metabolic realm. There ruminants, and especially cattle, develop their higher but dreaming-sleeping intelligence. The mesentery is the physical carrier of this intelligence which the cow, for instance, develops in performing the 'cosmic qualitative analysis' of the fodder and thereby comparing and composing the quality of the manure. In a higher sense the mesentery is the brain of the cow. As a sensory membrane it lines the inner sphere of the abdominal cavity and envelops all the organs within.

Going backwards in the cow and yet on a higher level than its sensory concentration in the head, we find this inner heaven, this wonderful mesentery which forms the serosa, the outer layer of the

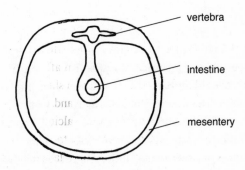

Vertical section of cow's abdominal region

metabolic organs. As a thin interwoven transparent membrane, it comes down from the spinal region of the abdominal cavity, envelops the intestines and goes back upwards again. So in between the spine region and the intestines, we have spread out a double membrane, a sensory organ which is sensitive to both sides. We really have an intensification of the skull/head function because this doubly active sensitive organ reflects and concentrates whatever the cow has perceived in analysing the composition of matter in the process of digestion. This happens mainly when the blood stream as carrier of the nourishing forces coming from the fodder flows up into the head, into the horn bone. There it cannot get any further; it gets dammed by the horns. The horn reverts the bloodstream back into the belly and then the peritoneum senses the result of this fodder analysis. It is within this region that the cow develops her consciousness and her powerful astrality. We now envelop the dandelion flowers in the mesentery and we have an inner and an outer sensitive preserving layer towards the flowers and an outer sensitive perceiving layer towards the elements and working ethers, to which we expose the wrapped up balls, when we bury them during wintertime. So I think the mesentery of the cow as part of the peritoneum is the ultimate we can find in nature to make this preparation. It really serves as a preserving, transmitting, concentrating and preserving organ.

Valerian

When we finally come to the third of the second group of preparations, the valerian, there is no appropriate animal organ left in nature that could further enhance or concentrate the working of this unique herb. The diluted juice of valerian by itself has the faculty to create a sheath, wherever it is sprayed. As I understand, it is a sheath which consists of condensed warmth ether. It opens towards the spiritual, protects against influences of the natural environment and preserves the workings of its brother and sister preparations when we spray it over the compost heap. The compost or dung heap is enveloped by a warmth-sheath and thus becomes an individualized realm of well organized decomposition of organic residue and humus formation.

When, in concluding, we look at the different sheaths we can comprehend their specific function both as being sense organs towards the outside world and concentration and preservation organs towards their inner content. Their sense activity is a most essential part of the preparation process. It is the condition of what becomes this powerful concentration of forces in the humus-like substance of the preparations. But are the preparations themselves sense organs? They are, in the same sense as a germinating plant senses the sunlight and all the workings of the cosmic periphery. As germ they are sense organs in the physical mineral world. They sense the progressing cosmos and thus have the potential to overcome the physical, dead material in nature. They are as it were a resurrection substance. They do not perceive and concentrate into an image, like an ordinary sensory organ, something which is there which has already evolved. The preparations draw in by a kind of perception, forces and substances from the cosmos, which endow the earth with the power of a new becoming. They are germinating carriers of spiritualizing evolving forces in the decaying physical-mineral world.

References

Chapter 1

1 Rudolf Steiner's report of June 20, 1924, after the Agriculture Course. In Steiner, *Agriculture.*
2 See *Tomorrow's Agriculture,* lecture on 'The History of Agriculture' by Manfred Klett.
3 See Manfred Klett, *Agriculture as an Art.*

Chapter 2

1 See Goethe, *Naturwissenschaftliche Schriften* [scientific writings], part 2.

Chapter 3

1 See John Wilkes, *Flowforms.* See also www.anth.org/virbelaflowforms
2 Steiner, *Man as Symphony of the Creative Word.*
3 Manfred Klett, *Agriculture as an Art.*

Chapter 4

1 Rudolf Steiner's 'Letters to Members,' published as *Anthroposophical Leading Thoughts.*
2 Pettersson, B. and Wistinghausen, E. von, *Bodenuntersuchungen zu einem langjährigen Feldversuch in Järna, Schweden.*
3 For information see Joachim Raupp (ed.) *Main Effects of Various Organic and Mineral Fertilization.*

Chapter 5

1 Rudolf Steiner, *Agriculture,* lecture 5.
2 Rudolf Steiner, *Anthroposophical Leading Thoughts,* see letter 1.

Chapter 6

1 Rudolf Steiner, *Man as Symphony of the Creative Word.*
2 Rudolf Steiner, *The Four Seasons and the Archangels.*

Chapter 7

1 Rudolf Steiner's report of June 20, 1924, after the Agriculture Course. In Steiner, *Agriculture.*
2 Rudolf Steiner, 'The Christ Impulse in the Course of History,' lecture given in Locarno, September 19, 1911.

Bibliography

Castellitz, K. *Life to the Land,* Lanthorn Press 1980. *Filled with experiences gleaned from a life time's work with the preparations.*

Corrin, George, *Handbook on Composting and the Preparations,* BDAA, UK1995. *Written by a long-time biodynamic consultant in the UK.*

Giono, Jean, *The Man who Planted Trees,* Peter Owen, London 1989.

Goethe, Johann Wolfgang von, *Naturwissenschaftliche Schriften,* Artemis, Zurich 1966.

Grotzke, Heinz, *Biodynamic Greenhouse Management,* BDF&GA, USA 1998. *Grotzke draws on over forty years of experience to provide a comprehensive perspective as well as practical tips on soil blends, biodynamic preparations, water, light, sanitation and cuttings.*

Keyserlingk, Adalbert Count, *Developing Biodynamic Agriculture,* Temple Lodge Press, Forest Row 1999. *Count Keyserlingk, the son of the hosts, was present when the agriculture course was given on the Koberwitz estate in Silesia. This book includes Keyserlingk's personal experience of Rudolf Steiner at work, his reflections on practical research and experimentation and descriptions of the preparations.*

Klett, Manfred, *Agriculture as an Art, the Meaning of Man's Work on the Soil,* IBIG, Sussex 1992.

— and Edmunds, Francis, *Tomorrow's Agriculture ... Are We Meeting the Challenges?* IBIG, Sussex 1986.

—, Moore, Rose and Carnegie, Andrew, *Growing Together ... Why Should We Bother?* IBIG, Sussex 1990.

— and Rapsey, Tim, *Dying and Becoming, Man's Path to a New Communion with Nature,* IBIG, Sussex 1990.

— and Spence, Michael, *Building Stones for Meeting the Challenges,* IBIG, Sussex 1988.

Koepf, Herbert, *The Biodynamic Farm,* Anthroposophic Press, NY 1989. *Based on years of research, Koepf demonstrates the practical feasibility of the biodynamic approach and how biodynamics protects and nurtures the soil, improving food quality so that farmers, gardeners and consumers can live in harmony with the environment.*

—, *Biodynamic Sprays,* BDF&GA, USA 1971. *This introductory booklet gives a basic description of how the biodynamic sprays are made, how they influence cultivation, and how to use them.*

—, *Research in Biodynamic Agriculture – Methods and Results,* BDF&GA, USA 1993. *A summary of research up to1992.*

König, Karl, *Earth and Man,* BDF&GA, USA 1982. *These lectures, given to farmers and gardeners by the founder of the Camphill Movement, contain thought-provoking insights into the nature of living organisms and a foundation for understanding the biodynamic preparations.*

Pettersson, B. and Wistinghausen, E. von, *Bodenuntersuchungen zu einem langjährigen Feldversuch in Järna, Schweden,* [ground tests on long-term field trials in Järna, Sweden] Forschungsring für biologisch-dynamische Wirtschaftweise, Darmstadt 1977.

Pfeiffer, Ehrenfried, *The Biodynamic Treatment of Fruit Trees, Berries and Shrubs,* BDF&GA, USA 1957. *Basic principles and practical guidance on growing top and soft fruit. It describes measures to take in order to develop a pest-free orchard without the use of chemicals.*

Raupp, Joachim (ed.) *Main Effects of Various Organic and Mineral Fertilization on Soil Organic Matter Turnover and Plant Ground,* Biodynamic Research Institute, Darmstadt 1995.

Remer, Nikolaus *Laws of Life in Agriculture,* BDF&GA, USA 1995. *Based on more than twenty-five years of farm research and experience as a biodynamic consultant in northern Germany, Remer offers detailed practical advice and far-reaching insights into biodynamic farming.*

Sattler, F. and E. vonWistinghausen, *Biodynamic Farming Practice,* BDAA, UK1992. *The definitive manual on biodynamics. This is a thorough and practical textbook describing in detail scientifically proven biodynamic techniques developed over many years.*

Schilthuis, Willy, *Biodynamic Agriculture,* Floris Books, Edinburgh 2004. *A handy pocket-book full of practical information on all aspects of biodynamic farming.*

Steiner, Rudolf, *Agriculture,* BDF&GA, USA 1993. *With this remarkable series of lectures given in Koberwitz, Silesia, in June 1924, Rudolf Steiner founded biodynamic agriculture. They contain profound insights into farming, the plant and animal world, the nature of organic chemistry and the influences of heavenly bodies.*

—, *Anthroposophical Leading Thoughts,* Rudolf Steiner Press, London 1985.

—, 'The Christ Impulse in the Course of History,' published in *Anthroposophical Quarterly,* Vol 12, No 13, Autumn 1967.

—, *An Outline of Esoteric Science,* Anthropsophic Press, NY1997.

—, *The Four Seasons and the Archangels,* Rudolf Steiner Press, London 1996.

—, *Knowledge of the Higher Worlds. How is it Achieved?* Rudolf Steiner Press, London 1969.

—, *Man as Symphony of the Creative Word* (twelve lectures given in Oct–Nov 1923), Rudolf Steiner Press, London 1991.

—, *Philosophy of Spiritual Activity,* Rudolf Steiner Publications, NY 1980.

Thun, Maria, **Gardening for Life,** Hawthorn Press, Stroud 1999. *A practical introduction to the holistic approach of biodynamic gardening, sowing, planting and harvesting. This beautifully structured book demystifies the relationship between plants and the cosmos and explains all you need to know to find the most favourable time and circumstance for everything from planting, to harvesting and preserving. Maria Thun is a well-known gardener and author on biodynamic techniques.*

Wilkes, John, *Flowforms,* Floris Books, Edinburgh 2003.

Wistinghausen, C. von, *et al.* **The Biodynamic Spray and Compost Preparations – Production Methods,** BDAA, UK2000. *A practical guide on making the preparations.*

—, **The Biodynamic Spray and Compost Preparations – Directions for Use,** BDAA, UK2002. *A practical handbook on how to use the preparations.*

Wright, Hilary, **Biodynamic Gardening for Health and Taste,** Mitchell Beazley, London 2003. *A colourfully illustrated introduction to biodynamics. It offers a great deal of historical and philosophical background information and while focussed primarily on the biodynamic garden, it explores the cultural/artistic context of biodynamics.*

Regular Journals

Star and Furrow, twice a year, UK
Biodynamics, six times a year, USA
Harvest, quarterly, New Zealand
Newsleaf, quarterly, Australia
Lebendige Erde, six times a year, Germany
Kultura, quarterly, Sweden
Available from the relevant Associations.

Useful contact addresses

United Kingdom

Biodynamic Agricultural Association (BDAA)

Painswick Inn Project, Gloucester Street, Stroud, GL5 1QG
Tel 01453-759 501
Fax 0845-345 8474
office@biodynamic.org.uk
www.biodynamic.org.uk
In existence since 1928, the Biodynamic Agricultural Association provides a membership service open to everyone interested in learning about, working with, developing or supporting this unique form of organic husbandry. A journal The Star and Furrow published twice yearly is available free to members together with a quarterly newsletter. A wide range of books is on sale and a member's library exists, containing many out of print books.
Contact the BDAA for details of: Regional Preparation Making (day introduction in spring and autumn); Training Opportunities.

Demeter Seeds

Stormy Hall Seeds, Botton Village, Danby, Whitby, N Yorks YO21 2NJ
stormy.hall.botton@camphill.org.uk
A wide range of biodynamically grown vegetable herb and flower seeds are available by mail order, an increasing number of UK origin.

Demeter Certification

Demeter is a symbol recognized throughout the world to describe biodynamic produce grown to strict production standards
Contact: Demeter Scheme Co-ordinator, 17 Inverleith Place, Edinburgh, EH3 5QE
Tel 0131-624 3921
Fax 0131-476 2996

Demeter International

Sekretariat, Kraaybeekerhof, Postbus 17, 3970 AA Driebergen, Netherlands
Tel +31-343-51 29 25
Fax +31-343-51 69 43
kraayhof@worldaccess.nl

Biodynamic Associations

Australia

Biodynamic AgriCulture Australia, P.O. Box 54, Bellingen, NSW 2454
Tel 02-6655 0566
Fax 02-6655 8551
bdoffice@biodynamics.net.au
www.biodynamics.net.au

Brazil

Instituto Biodinamico, Rua Amando de Barros 321, 18603-970, Botucatu SP
Tel 014-822 5066
Fax 014-822 5066
ibd@laser.com.br

Canada
Society for Biodynamic Farming &
Gardening in Ontario,
R.R #4 Bright, Ontario NOJ 1BO
Tel/Fax 519-684-6846

Egypt
Egyptian Biodynamic Association,
Heliopolis, El Horreya,
PO Box 2834, Cairo
Tel 02-280 7994
Fax 02-280 6959
ebda@sekem.com

France
Movement de Culture Biodynamique
5 place de la Gare, 68000 Colmar
Tel 03-8924 3641
Fax 03-8924 2741
Biodynamis@wanadoo.fr

Germany
Forschungsring für biologisch-
dynamische Wirtschaftsweise e.V.,
Brandschneise 2, 64295 Darmstadt
Tel 06155-84 12 41
Fax 06155-49 84 69 11

India
Biodynamic Agriculture
Association of India
31 Signals Vihar, Mhow, MP 453442
Tel 07324-746 64
Fax 07324-731 33

Ireland
Biodynamic Agricultural
Association in Ireland
The Watergarden, Thomastown,
Co.Kilkenny
Tel 056-54214
bdaai@indigo.ie

Italy
Assoziazione per l'Agricultura
Biodinamica
Via Vasto 4, 20121 Milano
Tel 02-2900 2544
Fax 02-2900 0692

Netherlands
Vereniging voor Biologisch
Dynamische Landbouw
Postbus 17, 3970 AA Driebergen
Tel 0343-531 740
Fax 0343-516 943
bd.vereniging@ecomarkt.nl

New Zealand
The Biodynamic Farming and
Gardening Association in N.Z. Inc,
PO Box 39045, Wellington Mail
Centre
Tel 04-589 5366
Fax 04-589 5365
biodynamics@clear.net.nz

South Africa
The Biodynamic Association of
Southern Africa,
PO Box 115, Paulshof 2056
Tel 011-803 7191
Fax 011-803 7191

Sweden
Biodynmiska Föreningen,
Skillebyholm, 15391 Järna
Tel 08-5515 1225
Fax 08-5515 1227

Switzerland
Landwirtschafliche Abteilung am
Goetheanum,
Hügelweg 59, 4143, Dornach
Tel 061-706 4212
Fax 061-706 4215
landw.abteilung@goetheanum.ch

USA
Biodynamic Farming and
Gardening Association Inc,
25844 Butler Road, Junction City,
OR 97448
Tel 541-998 0105
Fax 541-998 0106
biodynamics@aol.com
www.biodynamics.com

Index

Photographic Acknowledgments

Bernard Jarman: pp. 25, 75
Richard Swann: pp. 28, 79.

Results from the Biodynamic Sowing and Planting Calendar

Maria Thun

For over forty years, Maria Thun has been researching optimum days for sowing, pruning and harvesting various plant crops. Here, collected together for the first time, are results from this work, showing the influence of the rhythms of sun, moon, and planets on plant growth.

This book shows that if farmers and gardeners link their work into these cosmic rhythms, the quality of their produce is markedly increased.

Avoiding unsuitable days is shown to help prevent crop damage through disease and pests. Methods of fertilizing and spraying have been developed which further enhance produce, allowing a sustainable and ecologically balanced agriculture.

Includes sections on the stars, the soil, composting and manuring, weeds and pests, as well as growing cereals, vegetables, herbs, fruit and vineyards.

Floris Books